W0016787

GEOTECHNICAL SPECIAL PUBLICATION NO. 97

INNOVATIONS AND APPLICATIONS IN GEOTECHNICAL SITE CHARACTERIZATION

PROCEEDINGS OF SESSIONS OF GEO-DENVER 2000

SPONSORED BY
The Geo-Institute of the American Society of Civil Engineers

August 5-8, 2000
Denver, Colorado

EDITED BY
Paul W. Mayne
Roman Hryciw

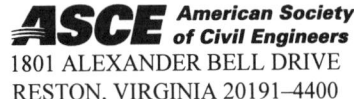 American Society of Civil Engineers
1801 ALEXANDER BELL DRIVE
RESTON, VIRGINIA 20191–4400

Abstract: In situ testing provides a direct means for evaluating natural and man-made ground conditions for geotechnical site characterization. Soils and rocks can be subjected to loading under ambient anisotropic stress states, complete with effects of inherent fabric, aging, cementation, sensitivity, fines content, and structure. The variety of field tests are particularly useful for designing different types of foundation systems, assessing soil liquefaction potential, and verifying soil improvement works. Many in situ tests now provide continuous and real time data with multiple independent readings taken within a single sounding. Seventeen technical papers are compiled in this geotechnical special publication under the theme of *Innovations & Applications in Geotechnical Site Characterization*. Recent advances have developed in the interpretation of in situ penetration test data from cone soundings (CPT, CPTu), flat plate dilatometer (DMT), pressuremeter tests (PMT), soil borings (SPT), as well as geophysical techniques (CHT, DHT). In addition, new devices have been invented to provide higher stratigraphic resolution and to quantify specific soil parameters and properties.

Library of Congress Cataloging-in-Publication Data

Geo-Denver 2000 (2000 : Denver, Colo.)
    Innovations and applications in geotechnical site characterization : proceedings of sessions of Geo-Denver 2000, August 5-8, 2000, Denver, Colorado / edited by Paul W. Mayne, Roman Hryciw.
        p. cm. – (Geotechnical special publication ; no. 97)
    Includes bibliographical references and index.
    ISBN 0-7844-0505-0
    1. Engineering geology—Congresses. 2. Soils—Testing—Congresses. 3. Building sites—Evaluation—Congresses. I. Mayne, Paul W. II. Hryciw, Roman D. III. American Society of Civil Engineers. Geo-Institute. IV. Title. V. Series.

TA703.5. G417 2000a
624.1'51—dc21

                                                    00-034252

Geotechnical Special Publications

Foreword

In situ tests provide stratigraphic delineation and direct information on geomaterial properties under actual field conditions, current states of stress, and inherent fabric. By contrast to laboratory testing of recovered samples, in situ testing often provides real-time and continuous profiling data for the immediate assessment of soils and rocks under anisotropic stresses and natural environments. Many recent developments have been seen in equipment, procedures, and interpretation of measurements in the use of in situ testing for geotechnical site characterization, particularly with respect to foundation systems, embankment construction, highways, environmental concerns, ground modification, and soil liquefaction hazard assessment.

This technical publication provides a look at the most recent novel devices and improved means of assessing ground conditions. Papers were prepared by both researchers and practitioners for presentation at the geotechnical specialty conference held by the ASCE Geo-Institute in Denver, Colorado during August 5-8, 2000. All of the papers underwent a peer-review process similar to that required by the ASCE *Journal of Geotechnical & Geoenvironmental Engineering* with a minimum of two independent consensus reviews. Each paper was revised to comply with the review comments prior to publication. All papers are open for technical discussion in the *Journal of Geotechnical & Geoenvironmental Engineering* and are eligible for nomination for ASCE awards.

A total of 17 papers survived the critical peer-review process. These papers are organized within five subheadings under the theme of the sessions on *Innovations & Applications in Geotechnical Site Characterization*, including: (1) in situ testing for liquefaction evaluation; (2) novel and innovative devices; (3) data processing and interpretation; (4) ground improvement evaluation by in situ tests; and (5) field tests for rock formations.

The editors would like to acknowledge the efforts of the important yet anonymous technical reviewers in providing the necessary critiques of the papers. In addition, the assistance and guidance of Harold W. Olsen, editor-in-chief of JGGE, Maria Marshall of Arizona State University, and Donna Dickert of ASCE are greatly appreciated.

The Editors

Paul W. Mayne, PhD, P.E.
Civil & Environmental Engineering
Georgia Institute of Technology
790 Atlantic Drive
Atlanta, GA 30332-0355
Email: pmayne@ce.gatech.edu

Roman D. Hryciw, PhD
Civil & Environmental Engineering
University of Michigan
2366 G.G. Brown Building
Ann Arbor, MI 48109-2125
Email: romanh@umich.edu

Contents

Section IV: Ground Improvement Evaluation by In Situ Tests

Section V: Field Testing for Rock Formations

Indexes

LIQUEFACTION RESPONSE OF SOILS IN MID-AMERICA EVALUATED BY SEISMIC CONE TESTS

By James A. Schneider[1] and Paul W. Mayne[2]

ABSTRACT

The Mid-America earthquake region is recognized as containing significant seismic hazards from historically-large events that were centered near New Madrid, MO in 1811 and 1812 and Charleston, SC in 1886. Methods for evaluating ground hazards as a result of soil liquefaction and site amplification are needed in order to properly assess risks and consequences of the next seismic event in these areas. In-situ tests provide quick, economical, and practical means for these purposes. For this research effort, seismic piezocone penetration tests (SCPTu) have been performed at a number of sites in the heart of the Mid-America earthquake regions. Many of these sites have already been associated with liquefaction features such as sand dikes, sand boils, or subsidence, observed during geologic and paleoseismic studies. Data collected at these sites have been analyzed under current methodologies to assess the validity and internal consistency of empirical relations developed for Chinese, Japanese, and Californian interplate earthquakes when applied to historical Mid-American earthquakes. Validation of extrapolation of cyclic resistance curves to high cyclic stress ratio values will be considered. Lower bound estimates of moment magnitude (M_w) from parametric studies on SCPTu data indicate an earthquake event of magnitude greater than 7.0 would have been necessary to induce soil liquefaction at the sites studied.

INTRODUCTION

It is now recognized that several of the largest historical earthquake events in the United States occurred in the New Madrid, MO area during 1811 and 1812, and in Charleston, SC in 1886. Large events prior to these times are also acknowledged. The New Madrid

---

[1] Staff Engineer, GeoSyntec Consultants, 1100 Lake Hearn Drive, Suite 200, Atlanta, GA 30342, e-mail: jamess@geosyntec.com

[2] Professor, Georgia Institute of Technology, School of Civil & Environmental Engineering, Atlanta, GA 30332-0355, e-mail: paul.mayne@ce.gatech.edu

series of 1811-1812 consisted of over 200 separate seismic events, which would have created an equivalent single event with a moment magnitude (M_w) of about 8.3 (Johnston & Schweig, 1996). The three largest individual events of the series were estimated to have moment magnitudes estimated at 7.9, 7.6, and 8.0 on December 16, 1811, January 23, 1812, and February 7, 1812 respectively (Johnston & Schweig, 1996). The Charleston, SC earthquake consisted of a single event on September 1, 1886, with a M_w estimated at 7.0 (Stover & Coffman, 1993).

Ongoing research on the magnitude, attenuation, and recurrence of earthquake events in Mid-America has led to the increased awareness of the potential for serious ground failures in the New Madrid, MO seismic zone and Charleston, SC earthquake region. Strong ground motions can lead to injury and death from damaged structures, primarily from the collapse of buildings and bridges. Site amplification and liquefaction-induced ground failures may increase the severity of earthquake effects. Large lateral and vertical movements will rupture pipelines and utilities, crippling lifeline facilities needed to provide aid and relief to the injured.

It will be desirable to evaluate the response of soils to earthquake shaking and potential for liquefaction in an expedient and cost effective manner in the Central and Eastern United States (CEUS). However, the evaluation of liquefaction response of soils is complicated in Mid-America due to the:

- deep vertical soil columns (600 m to 1400 m) of the Mississippi River Valley and Atlantic Coastal plain;
- infrequency of large events needed for calibration of models and analysis techniques (most recent sever event, $M_w > 6.5$, more than 100 years ago);
- uncertainty associated with the mechanisms and subsequent motions resulting from intraplate earthquakes (e.g., California earthquakes are interplate events).

SEISMIC PIEZOCONE TEST

To obtain parameters for engineering analysis and model studies, field test data are necessary. The seismic piezocone penetrometer is an electronic probe that rapidly provides four independent parameters to assess the subsurface profile with depth at an individual site. Figure 1 presents data from a seismic piezocone sounding in West Memphis, AR, including tip resistance corrected for pore pressure effects (q_t; Lunne et al., 1997), sleeve friction (f_s), penetration porewater pressure measurement measured behind the tip (u_2), and shear wave arrival time (t_s). The arrival time is incorporated into a pseudo interval analysis method (Campanella et al., 1986) for determination of shear wave velocity (V_s).

With regards to liquefaction evaluation, the individual recordings from seismic piezocone penetration tests (SCPTu) can be valuable in evaluating input parameters as illustrated by Figure 2. Specifically, the readings are processed to obtain:

- Direct measure of small strain shear stiffness ($G_{max} = \rho \cdot V_s^2$);
- Soil type and stratigraphy (q_t, $FR=f_s/q_t \cdot 100$, u_2);

- Depth of water table in sands (u_2);
- Liquefaction susceptibility from direct analysis (q_c and V_s);
- Estimations of properties for rational analysis (ϕ', D_r, OCR, K_o).

The additional dynamic soil properties of peak particle velocity (PPV or $\dot{u}$), and strain level (γ_s = PPV / V_s) can be determined from the shear wave velocity and geophone output (Figure 3).

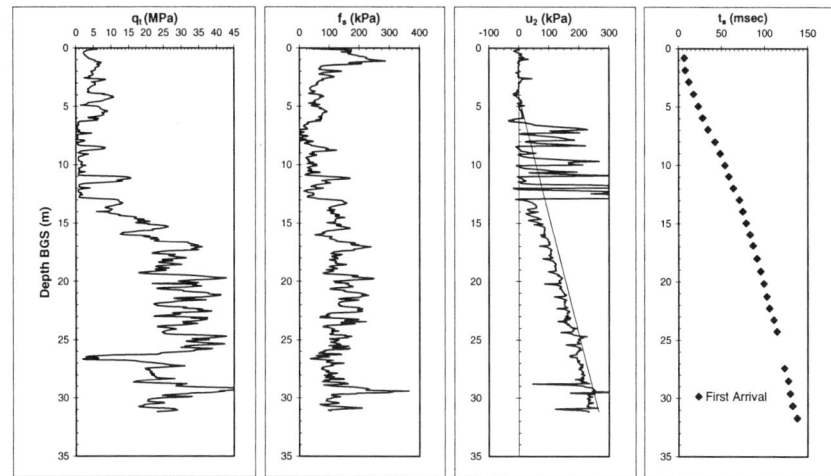

Figure 1. Raw Data from Seismic Piezocone in West Memphis, AR (MEMPH-K)

SEISMIC GROUND MOTIONS IN MID-AMERICA

Before an earthquake analysis can be performed, critical ground motion parameters must be selected. An assessment of ground motion hazards is difficult in the Mid-America earthquake region due to the lack of strong earthquakes in recent historical times (t ≈ 100+ years), and lack of recorded data from the limited events that have occurred. A stochastic ground motion model has been under development for the Central and Eastern United States (CEUS), and attenuation relationships for rock sites have been formed using this model (e.g., Toro et al., 1997). Synthetic ground motions based on a representative stiffness profile of the Mississippi River Valley deep soil column are under development for Mid-America (Herrmann & Akinci, 1999).

For this study, maximum horizontal acceleration (a_{max}) and Arias intensity (I_h) were determined using the Herrmann & Akinci (1999) ground motions program. The output

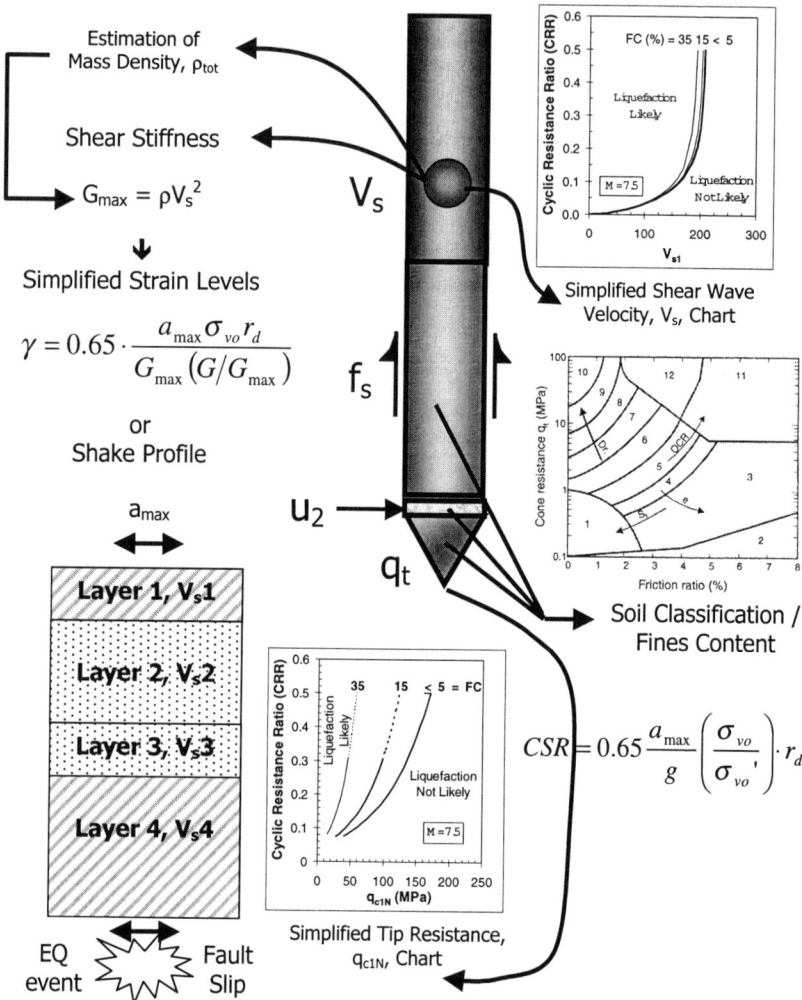

Figure 2. Seismic Piezocone Parameters used for Earthquake Analysis of Soil

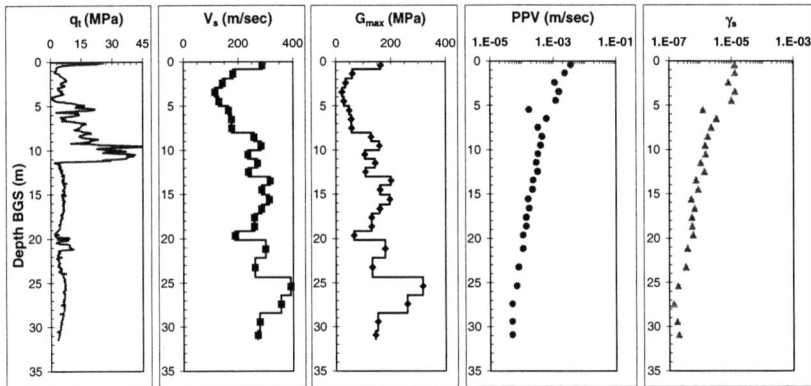

Figure 3. Dynamic Properties Determined from Seismic Piezocone Sounding at Shelby Farms, Shelby County, TN

of this program was primarily a function of the anticipated depth of the soil column, hypocentral distance to the site and moment magnitude. To determine hypocentral distance as a function of Joyner-Boore distance or epicentral distance the hypocentral depth is necessary. The hypocentral depth was assumed to be 9.3 km in the NMSZ and 10.9 km in Charleston, SC based on the work of Toro et al., (1997). Five ground motion models are available with the Herrmann & Akinci (1999) program, differing in spectral source, wave propagation model, and soil conditions. The model used in this study combined USGS 150-bar spectral scaling (Frankel et al., 1996), with Atkinson and Boore (1995) wave propagation, and a generic deep stiffness profile for the NMSZ (Herrmann et al., 1999).

LIQUEFACTION ANALYSIS OF SOILS IN MID-AMERICA

Since the effects of structure, aging, cementation, and strain history generally cannot be replicated in laboratory specimens of granular materials, the use of in-situ testing results and field performance data has become a popular means of assessing liquefaction susceptibility. In-situ test parameters at sites where surface manifestations of liquefaction were or were not evident have been compared to evaluate cyclic soil resistance. Databases consisting predominantly of sites from China, Japan, and California are available for the Standard Penetration Test (SPT; e.g., Seed et al., 1983), cone penetration test (CPT; e.g., Olson & Stark, 1998), flat dilatometer test (DMT; e.g., Reyna & Chameau, 1991), and shear wave velocity (V_s; Andrus et al., 1999). Analyses by these methods are considered as direct methods for liquefaction assessment of soils. It should be noted that these databases are applicable to Holocene deposits. The soils evaluated in this study consisted of Holocene deposits at sites west of the Mississippi River, with older Wisconsin deposits in the Memphis, TN area.

Evaluations based on two analysis procedures using SCPTu data from the Mid-America region are presented:
1. Cyclic stress based procedures for normalized CPT tip resistance (Robertson & Wride, 1998) and stress corrected shear wave velocity (Andrus et al., 1999);
2. Arias intensity methods for cone tip resistance adapted from the work of Kayen & Mitchell (1997).

A brief discussion of each analysis procedure will be presented, with more detail on each method given in Schneider & Mayne (1999) and the associated references presented above.

Cyclic Stress Based Procedures

Simplified cyclic stress based procedures require the three input parameters of (1) cyclic stress ratio (CSR); (2) normalized in-situ test parameter for which a liquefaction case history databases exists; and (3) cyclic resistance ratio (CRR).

The cyclic stress ratio is a function of the anticipated earthquake and is expressed as:

$$CSR = \frac{\tau_{avg}}{\sigma_{vo}'} \approx 0.65 \cdot \left(\frac{a_{max}}{g}\right) \cdot \left(\frac{\sigma_{vo}}{\sigma_{vo}'}\right) \cdot r_d \tag{1}$$

where a_{max} is the maximum horizontal surface acceleration (in g's), σ_{vo} is the total vertical stress at the depth of concern, σ_{vo}' is the effective vertical stress at the depth of concern, and r_d is a stress reduction coefficient. The magnitude and depth dependent stress reduction coefficient (r_d) as presented by Idriss (1999) was used in this study.

Normalized cone tip resistance (q_{c1N}) as expressed in Equation 2 was utilized for this study to maintain consistency with the existing liquefaction databases and previously proposed CRR curves (Robertson & Wride, 1998).

$$q_{c1N} = q_c / (\sigma_{vo}')^n \tag{2}$$

where $n = 0.5$ in clean sands, $n = 0.75$ for silty sands, and q_c and σ_{vo}' are in atmospheres. This CPT tip resistance normalization is not used for soils with fines content greater than 35 percent, but all critical layers this study contained less than 35 percent fines. Overburden stress corrected shear wave velocity (V_{s1}) is expressed as (Robertson et al, 1992):

$$V_{s1} = V_s / (\sigma_{vo}')^{0.25} \tag{3}$$

where V_s is in m/s and σ_{vo}' is in atmospheres. The cyclic resistance ratio (CRR) is an empirical relationship between a stress normalized in-situ test parameter (e.g., q_{c1N}, V_{s1}) and a soils resistance to cyclic stresses imposed by an earthquake event representing a factor of safety of unity (FS=1). Since the magnitude of an earthquake will effect the cyclic resistance and is not incorporated into the CSR (Eq. 1), the cyclic resistance ratio is normalized to a moment magnitude (M_w) earthquake of 7.5 ($CRR_{7.5}$) using a magnitude scaling factor (MSF). The Idriss (1999) magnitude scaling factors expressed in the following equation were used in this study:

$$MSF = 31.9 \ (M_w)^{-1.72} \tag{4}$$

The cyclic resistance ratio curves examined in this study were based on the recommendations of NCEER (1997). For normalized cone tip resistance (q_{c1N}; Eq. 2) the $CRR_{7.5}$ is expressed as (Robertson & Wride, 1998):

$$\text{if } 50 \le (q_{c1N})_{cs} < 160 \qquad CRR_{7.5} = 93 \cdot \left(\frac{(q_{c1N})_{cs}}{1000} \right)^3 + 0.08 \tag{5a}$$

$$\text{if } (q_{c1N})_{cs} < 50 \qquad CRR_{7.5} = 0.83 \left(\frac{(q_{c1N})_{cs}}{1000} \right) + 0.05 \tag{5b}$$

where $(q_{c1N})_{cs}$ is a clean sand equivalent normalized cone tip resistance. Since soils in this study are considered relatively clean sands, no adjustment to q_{c1N} (Eq. 2) was necessary. For stress corrected shear wave velocity (V_{s1}; Eq. 3), the $CRR_{7.5}$ is represented as (Andrus et al., 1999):

$$CRR_{7.5} = a \cdot \left(\frac{V_{s1}}{100} \right)^2 + b \cdot \left(\frac{1}{V_{s1}^* - V_{s1}} - \frac{1}{V_{s1}^*} \right) \tag{6}$$

where V_{s1}^* is the limiting upper value of V_{s1} for liquefaction occurrence, and a and b are curve fitting parameters equal to 0.022 and 2.8 respectively. The limiting value of shear wave velocity in relatively clean sands (FC $\le$ 5) of concern for this study has been estimated to be 215 m/s (Andrus et al., 1999).

It is anticipated that an event in Mid-America could result in cyclic stress ratios on the order of 1.0 or higher at close epicentral distances (Toro et al., 1997). Current field performance data are limited to CSR values typically below 0.4, with most data in the 0.1 to 0.2 range. Reconstituted specimens used in laboratory tests do not fully replicate soil fabric, which will lead to different interpretations of liquefaction resistance from laboratory test data than field observations. Advances in sampling of granular soils by freezing techniques allows in-situ soil fabric to remain relatively undisturbed prior to laboratory testing. The cyclic resistance of a deposit may be estimated using laboratory test results from frozen specimens, but an accurate estimate of K_o will be necessary for a reasonable assessment of field liquefaction resistance.

Field performance based CRR curves can be validated by comparison of laboratory based cyclic resistance from frozen specimens to in-situ test parameters taken adjacent to the sample location. A study by Suzuki et al (1995) presents field shear wave velocity and cone tip resistance data as compared to laboratory cyclic resistance from frozen specimens. Figures 4a and 4b display the data compared to q_{c1} and V_{s1} respectively. The Robertson & Wride (1998) curves match the average value of the data presented in Suzuki et al. (1995) study, but a number of points are misclassified. The uncertainty inherent when using simplified curves should be modeled using a conservative estimate of liquefaction resistance with respect to the field data. The engineer can then judge the factor of safety they are comfortable with based on experience and/or probabilistic methods.

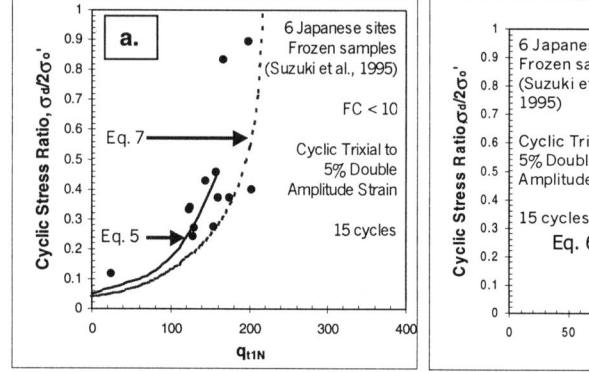 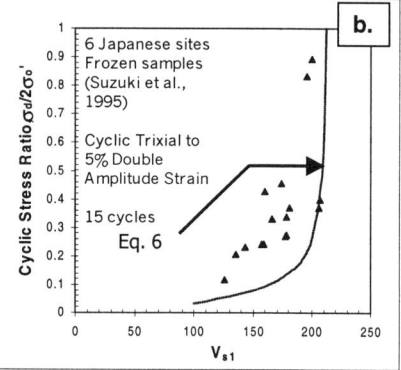

Figure 4. Comparison of CRR curves and Laboratory Frozen Specimens for (a) CPT Tip Resistance (b) Shear Wave Velocity

The format of the Andrus et al. (1999) CRR curve leads to an asymptotic value of shear wave velocity at high values of CSR. A similar form will be adapted for the CRR determined from CPT q_{c1N} data:

$$CRR = a \cdot \left(\frac{q_{c1N}}{350}\right) + \frac{b}{\left(q_{c1N}{}^{*} - q_{c1N}\right)} \qquad (7)$$

where a and b are curve fitting parameters equal to 0.7 and 9.33 respectively. The limiting value of normalized cone tip resistance in clean sands has been estimated at 230 using cyclic triaxial test data presented in Suzuki et al. (1995). The $-b / V_{s1}{}^{*}$ term from Equation 6 forces the V_s CRR curve through zero. Since it is accepted that the CRR does not pass through the origin (NCEER, 1997), the corresponding CPT term is left out of Equation 7. To validate this curve for field performance data, Figure 5 compares Equation 7 and Equation 5 using the Olson & Stark (1998) CPT liquefaction case history database. Equation 7 is more conservative than currently-recommended methods, but encompasses all of the sites in Figure 5 where liquefaction was evident.

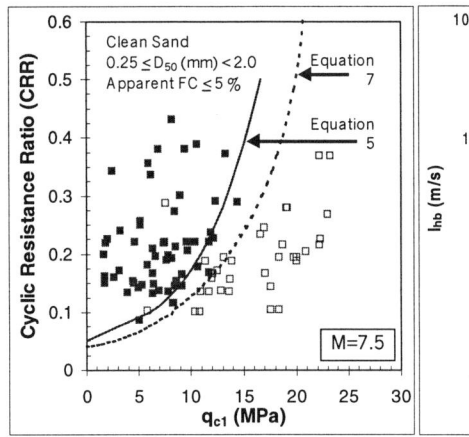

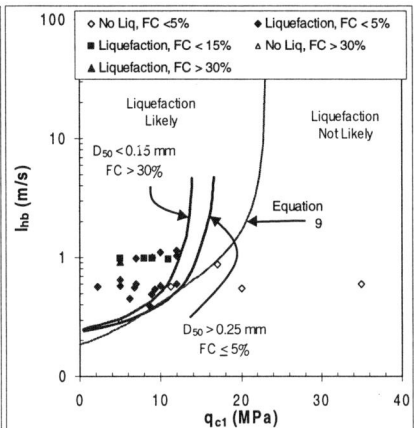

Figure 5. Comparison of CRR curves with CPT Field Performance data

Figure 6. Field Data Compared to Arias Intensity based Resistance Curves

Arias Intensity Based Procedures

A developing method for liquefaction analysis based on Arias intensity of earthquake records has the advantage that it does not require magnitude scaling factors. Although this method seems promising, lack of strong motion data in the Mid-America Earthquake region area leads to increased reliance on synthetic ground motion models. Arias intensity represents the cumulative energy per unit weight in a given direction that is absorbed by a set of single degree of freedom oscillators (Arias, 1970). Horizontal Arias intensity (I_h) can be calculated as the sum of Arias Intensity in the x- (I_{xx}) and y- (I_{yy}) directions as (Kayen & Mitchell, 1997):

$$I_h = I_{xx} + I_{yy} = \frac{\pi}{2g} \int_{.0}^{.t_o} a_x^{\,2}(t)dt + \frac{\pi}{2g} \int_{.0}^{.t_o} a_y^{\,2}(t)dt \tag{8}$$

where g is the acceleration due to gravity, $a_x(t)$ is a horizontal acceleration time history, and $a_y(t)$ is the horizontal acceleration time history in the direction perpendicular to $a_x(t)$.

Similar to the CSR from the Seed & Idriss (1971) simplified procedure, the Arias intensity will typically decrease with depth. Depending upon the depth where the time history was recorded and the depth of the liquefied layer, it may be necessary to apply a depth correction factor, r_b. The depth correction factors used in this study were as presented in Kayen & Mitchell (1997).

Simplified liquefaction resistance curves have been generated comparing Arias Intensity (I_{hb}) to penetration resistance [$(N_1)_{60}$ and q_{c1}] for field case histories where strong ground motion data have been readily available. These curves are based on limited field data from California (n=28), and thus Arias Intensity Resistance to liquefaction curves ($I_{hb}R$) are considered approximate. Considering Arias intensity field performance data for the CPT, cyclic stress-based field data for the CPT, and stress-based laboratory tests on frozen specimens compared to CPT tip resistance, the CPT-based liquefaction resistance curve should approach a vertical asymptote. To maintain consistency in analysis, this asymptote should be equal to that presented for cyclic stress-based procedures in clean sands: $q_{c1n}^* = 230$. Alteration of the curve fitting coefficients to account for differences between Arias intensity and cyclic stress based analyses, yields the following equation:

$$I_{hb}R = a \cdot \left(\frac{q_{c1N}}{350} \right)^2 + \frac{b}{\left(q_{c1N}^* - q_{c1N} \right)} \tag{9}$$

where a = 1.1 and b = 42.7. Figure 6 displays field performance data, the simplified curves from (Kayen & Mitchell, 1997), and the simplified curve presented in Equation 9, thus relating Arias intensity and normalized cone penetration tip resistance to liquefaction resistance. To maintain consistency with data presented in Kayen & Mitchell (1997), q_{c1N} is converted to the units of MPa for Figure 6. Both sets of curves match well with the limited field data, but Equation 9 approaches a more internally consistent asymptote at $q_{c1N}^* = 230$.

Selection of Critical Layers

To accommodate evaluation under a number of earthquake magnitude scenarios and liquefaction susceptibility frameworks, critical layers were selected for analysis using a procedure independent of earthquake magnitude and acceleration. A method was developed which combined selection of uniform layers for soil classification purposes (Olsen, 1994), and development of loosened and densified layers as a result of soil liquefaction (Youd, 1984).

In a study of historical California earthquakes, Youd (1984) discusses how expelled porewater from a liquefied deposit can be trapped beneath low permeability layers. This creates a loose layer below the low permeability cap due to the migration of porewater into that layer, and a densified layer below the loose layer due to migration of porewater out of that layer.

To assess uniform layers for classification purposes, Olsen (1994) used the rate of increasing tip resistance compared to effective overburden stress in layer selection techniques. Analysis of the data involved plotting CPT tip resistance and sleeve friction measurements compared to effective overburden stress on a log-log plot. Layers of constant soil type and consistency increase with effective confinement on a slope of $1/c$, where c is the stress exponent for normalization. Very dense, overconsolidated soils were determined to have a relatively vertical slope of log σ_{vc}' vs. log q_c, and thus a small value stress exponent, c, on the order of 0.15 or lower. Loose, soft soils were determined to have flatter slopes, and thus a *c-value* on the order of 1.0.

Seismic cone data collected at sites associated with an earthquake event was used to determine two critical layers; (1) loose granular layer with high liquefaction potential, and (2) a denser granular layer with lower liquefaction potential. Normalized layers were selected by evaluating cone tip resistance on a log scale plotted against effective vertical stress on a log scale. Figure 7 displays four channels of SCPTu data in a conventional manner, with the selected critical layers and associated parameters shown. Table 1 presents SCPTu parameters used for simplified analyses.

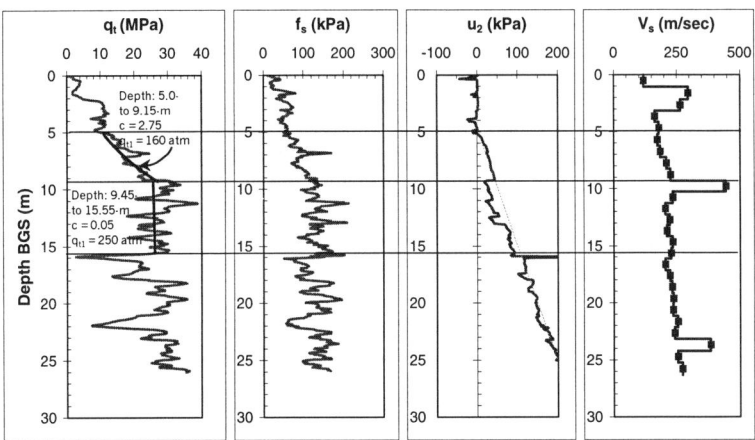

Figure 7. Normalization of Uniform Loose and Dense Sand Layers at Huey House, Blytheville, AR (based on standard plotting scales)

Parametric Study of Earthquake Scenarios

Seismic piezocone penetration tests at sites in Mid-America that have or have not shown surface evidence of liquefaction have been performed. Data from sites related to three historic earthquakes (Stover & Coffman, 1993), and three earthquakes dated by paleoliquefaction studies (Tuttle et al., 1998) have been analyzed by direct methods using simplified cyclic stress procedures, and simplified procedures utilizing Arias Intensity estimations (Schneider & Mayne, 1999). Due to space constraints, only data and analyses related to the New Madrid event of December 16, 1811 are presented.

A range of earthquake magnitude between 6.5 and 8.0 was studied, with associated accelerations and Arias Intensity estimated using the Herrmann & Akinci (1999) model. Table 2 displays the sites with the associated sounding I.D., estimated epicentral distance, and peak ground accelerations (PGA, a_{max}) associated with the range of earthquake magnitudes. To analyze a number of sites for various earthquake scenarios, critical layers independent of earthquake magnitude and acceleration were used (Table 1). Figure 8 displays the location of critical layers with respect to the cyclic resistance or Arias Intensity resistance curves. Sites with two critical layers have a solid symbol relating to a looser layer and an open symbol relating to a denser layer. Recommended curves by NCEER (1997) are shown as solid lines and the curves presented in equations 7 and 9 for q_{c1N} (CRR) and q_{c1N} (Arias) respectively are shown as dotted lines. Analyses for this study used curves presented in equations 6,7, and 9.

Table 1. Seismic Piezocone Parameters used for Simplified Analyses

Sounding	site lique-faction Y/N	water table m	top of layer m	layer thick-ness m	q_{t1}	c	$q_{t,avg}$ atm	q_{t1N}	$V_{s,avg}$ m/s	V_{s1} m/s	soil type[1]
MEMPH-G1	Y	6	6.8	2.5	100	3	160	148	172	165	SP
MEMPH-G2	Y	6	10.1	0.85	360	0.05	350	299	261	241	SP
SFSR-01	N	2	6.1	0.35	150	0.05	161	192	181	198	SP
MEMPH-H1	N	5	6.9	0.45	200	0.05	199	195	189	187	SP
MEMPH-H2	N	5	7.7	0.5	115	0.05	113	107	180	176	SP
SFOR-01	N	5.5	5.5	4.25	75	0	77	68	271	260	ML
YARB-01a	Y	4	15.35	2.5	20	5	213	167	234	207	SP
YARB-01b	Y	4	19.45	1.7	215	0.4	288	205	310	262	SP
3MS617-A1	Y	5.5	13	2.6	30	3.3	142	112	206	183	SP
3MS617-A2	Y	5.5	17.45	7.45	220	0.05	232	157	234	193	SP

q_{t1} - normalized tip resistance based on uniform layer using stress exponent c (Method shown in Fig. 7)
c - stress exponent for tip resistance normalization from graphical procedures (Method shown in Fig. 7)
$q_{t,avg}$ - tip resistance averaged over layer presented
q_{t1N} – overburden stress normalized average tip resistance (Eq. 2)
$u_{2,avg}$ - average penetration porewater pressure for layer taken behind the tip
$V_{s,avg}$ - shear wave velocity averaged over layer presented
V_{s1} – overburden stress corrected average shear wave velocity (Eq. 3)
[1] USCS Symbol

Table 2. Estimated Peak Ground Acceleration (g) for NMSZ 1811 Event

Moment Magnitude	Estimated Epicentral Distance (km)					
	Shelby Forest (SFOR) 40[a]	Yarbro Excavation (YARB) 60[b]	3MS617 (3MS617-A) 65[b]	Shooting Range (SFSR) 70[a]	Shelby Farms (MEMPH-G) 75[a]	Houston Levee (MEMPH-H) 90[a]
6.5	0.13	0.09	0.09	0.07	0.07	0.06
7.0	0.19	0.14	0.13	0.10	0.10	0.09
7.5	0.28	0.21	0.20	0.15	0.15	0.14
7.9	0.39	0.29	0.26	0.20	0.20	0.18

[a] 1000-m of soil over bedrock
[b] 600-m of soil over bedrock

Analysis of the data presented in Figure 8 is summarized in Table 3. Each critical layer is associated with a vertical line at a constant tip resistance or shear wave velocity representing a range of moment magnitudes studied. The point at which the vertical line crosses the resistance curve (CRR or I_hR) is deemed the critical lower bound magnitude for the analysis method and critical layer. If the intersection lies above $M_w=7.9$, but q_{c1N} is less than 230 or V_{s1} is less than 215 m/s, the critical layer may liquefy but at a magnitude higher than $M_w=7.9$. If q_{c1N} is greater than 230 or V_{s1} is greater than 215 m/s, it is considered that the critical layer will not liquefy regardless of moment magnitude.

The purpose of this study was to (1) check the internal consistency of current simplified liquefaction analyses when applied to sites in Mid-America; and (2) compare lower bound estimates of earthquake moment magnitude predicted by simplified charts to previous estimates of earthquake magnitude based primarily on intensity contours. It should be emphasized that these are lower bound estimates. The resistance curves are intended to represent a factor of safety of unity, and thus many of the sites may not be borderline cases with factors of safety much less than unity. Due to the size and abundance of liquefaction features in the areas studied, it is likely that the factor of safety may be less than unity. Additionally, since two other severe events took place within the same timeframe and general geographic area of the 1811 event, it is difficult to separate events which may have led to these liquefaction features a certain sites. Liquefaction features at the Yarbro excavation site and the 3MS617 site were

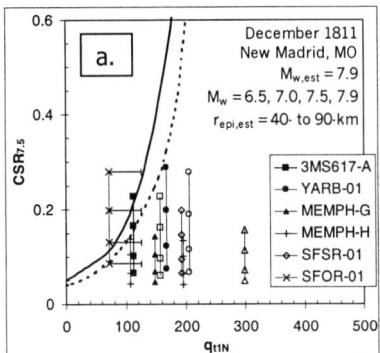

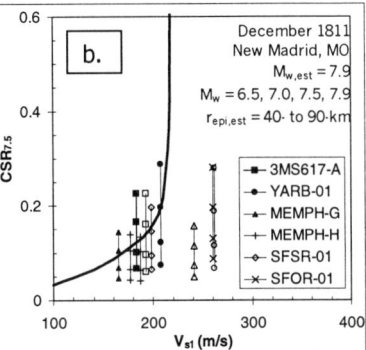

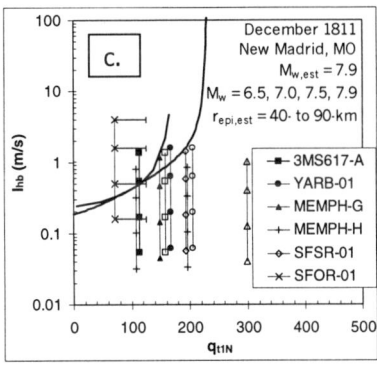

Figure 8. Liquefaction Analyses for NMSZ 1811 Event Using
(a) Cyclic Stress for CPT q_{c1N}
(b) Cyclic Stress for V_{s1}
(c) Arias Intensity for q_{c1N}

Table 3. Critical Lower Bound Magnitude

Sounding	$r_{epi,est}$	Site Liquefaction (Y/N)	Cyclic Stress q_{c1N}	Cyclic Stress V_{s1}	Arias Intensity q_{c1N}
MEMPH-G1	75	Y	> 7.9	7.5	7.7
MEMPH-G2	75	Y	N	N	N
SFSR-01	70	N	> 7.9	7.6	> 7.9
MEMPH-H1	90	N	> 7.9	7.6	7.7
MEMPH-H2	90	N	> 7.9	7.9	> 7.9
*SFOR-01**	*40*	*N*	*7.5*	*N*	*7.1*
YARB-01a	60	Y	8.1	7.6	7.7
YARB-01b	60	Y	N	N	> 7.9
3MS617-A1	65	Y	7.5	7.1	7.5
3MS617-A2	65	Y	> 7.9	7.4	7.7

* cemented loess layer which requires fines content correction, and possibly V_s cementation correction
N – will not liquefy, average parameter value greater than limiting asymptotic in-situ test parameter value

located at epicentral distances which may have been influenced by the events of January 23, 1812, and February 7, 1812.

DISCUSSION

From this study, the critical lower bound moment magnitude necessary to cause liquefaction based on tip resistance and cyclic stress based procedures ranges between 7.5 to greater than 7.9 for liquefaction sites, and greater than 7.9 for non-liquefaction sites. The critical M_w from V_{s1} cyclic stress based procedures ranges between 7.1 and 7.6 for liquefaction sites, and 7.6 to greater than 7.9 for non-liquefaction sites. The critical M_w from q_{c1N} Arias Intensity based procedures ranges between 7.5 and 7.7 for liquefaction sites, and 7.6 to greater than 7.9 for non-liquefaction sites. The Shelby Forest Site was left out of these ranges since it is a cemented silt loess that would require numerous correction factors for all three methods.

It can be inferred from the layering and consistency characteristics of soil deposits at liquefaction sites in the New Madrid seismic zone, that liquefaction resulted in loose sands near the surface from excess porewater pressures becoming trapped below low permeability clay or silt layers. While analyses from this study evaluated two separate layers per sounding, it is possible that liquefaction occurred throughout the depth of the deposit. This would result in the observed profile of a low permeability silty clay cap, over loose shallow sands, over densified deeper sands. Since databases used to generate cyclic resistance curves are based on the loosened post earthquake state of soil, the selected looser layers would be more appropriate for analyses under these frameworks.

The data available from the seismic piezocone penetration test provides the stiffness parameters necessary to perform a site-specific response analysis of ground motions, such as SHAKE. This would eliminate the empirical r_d and r_b stress reduction coefficients used in the simplified procedures. Strain levels for use in cyclic strain based methods can also be generated from a site-specific analysis. Assessment of analysis methods presented in this study utilizing equivalent linear and nonlinear models may provide additional insight into the liquefaction behavior of deposits profiled in this study.

CONCLUSIONS

The use of seismic piezocone tests for geotechnical site characterization provides four parameters that can be used for liquefaction assessment. Direct liquefaction analysis can be performed using cyclic stress based procedures for normalized cone tip resistance and stress corrected shear wave velocity to give redundancy in analysis, but inconsistencies in results leads to uncertainty. Magnitude-independent Arias intensity methods are also available for normalized cone tip resistance. These three methods were generally consistent for liquefaction analyses in Mid-America, but still resulted in a level of uncertainty that would require advanced analyses for borderline cases.

ACKNOWLEDGEMENTS

The authors were supported by the Mid-America Earthquake (MAE) Center, the National Science Foundation (NSF), and the U.S. Geologic Survey. Appreciation is expressed to the numerous researchers involved with the MAE Center and the Center for Earthquake Research and Information (CERI) who provided assistance with the many aspects of this project. Individuals at Georgia Tech and Illinois (UIUC) are thanked for assistance during the fieldwork portion of this research effort.

REFERENCES

Andrus, R.D., Stokoe, K.H., and Chung, R.M. (1999) Draft guidelines for evaluating liquefaction resistance using shear wave velocity measurements and simplified procedures, *NISTIR 6277*, NIST, Gaithersburg, MD, 121 pp.

Arias, A. (1970) A measure of earthquake intensity, *Seismic Design for Nuclear Power Plants*, MIT Press, Cambridge, MA.

Atkinson, G.M., and Boore, D.B. (1995) Ground-motion relations for eastern North America, *Bulletin of the Seismological Society of America*, 85, pp. 17-30.

Campanella, R.G., Robertson, P.K., and Gillespie, D. (1986) Seismic cone penetration tests, *Use of In-Situ Tests in Geotech Engineering*, ASCE GSP 6, pp. 116-130.

Frankel, A., Mueller, C., Barnhard, T., Perkins, D., Leyendecker, E.V., Dickman, N., Hanson, S., and Hopper, M. (1996) National Seismic-Hazard Maps: Documentation June 1996, *USGS Open-File report 96-532*.

Herrmann, R.B., and Akinci, A. (1999) Mid-America Ground Motion Models, Dept of Earth and Atm Sci, St Louis University, St. Louis, MO, 63103, 8 pp.

Herrmann, R.B., Ortega, R., and Akinci, A. (1999) Mid-America probabilistic hazard maps, Dept of Earth and Atm Sciences, St. Louis Univ, St. Louis, MO, 5 pp.

Idriss, I.M. (1999) An update of the Seed-Idriss simplified procedure for evaluating liquefaction potential, *Presentation Notes*, TRB'99 Workshop on New Approaches to Liquefaction Analysis, Washington D.C., January 10, 21 pp.

Johnston, A.C., and Schweig, E.S. (1996) The enigma of the New Madrid earthquakes of 1811-1812, *Annual Review of Earth and Planetary Sciences*, 24, pp. 339-384.

Kayen, R.E., and Mitchell, J.K. (1997) Assessment of liquefaction potential during earthquakes by Arias intensity, *Journal of Geotechnical and Geoenvironmental Engineering*, ASCE, 123 (12), pp. 1162-1174.

Lunne, T., Robertson, P.K., and Powell, J.J.M. (1997) *Cone Penetration Testing in Geotechnical Practice*, Blackie Academic & Professional, 312 pp.

NCEER (1997) Summary report, *Proceedings*, Workshop on Evaluation of Liquefaction Resistance, NCEER-97-0022, MCEER, Buffalo, NY, pp. 1-40.

Olsen, R.S. (1994) *Normalization and prediction of geotechnical properties using the cone penetration test (CPT)*, Report GL-94-29, USACOE, Vicksburg, 292 pp.

Olson, S.M., and Stark, T.D. (1998) CPT based liquefaction resistance of sandy soils, *Geotech. Earthquake Eng. and Soil Dynamics III*, ASCE GSP 75, pp. 325-336.

Reyna, F., and Chameau, J.L. (1991) Dilatometer based liquefaction potential of sites in the Imperial Valley, *Proceedings*, 2nd Int. Conference on Recent Advances in Geotech. Earthquake Eng. and Soil Dynamics, Vol. 1, St. Louis, pp. 385-392.

Robertson, P.K., Woeller, D.J., Kokan, M., Hunter, J., and Luternaur, J. (1992) Seismic techniques to evaluate liquefaction potential, *45th Canadian Geotechnical Conference*, Toronto, Ontario, October 26-28, pp. 5:1-5:9.

Robertson, P.K., and Wride, C.E. (1998) Evaluating cyclic liquefaction potential using the cone penetration test, *Canadian Geotechnical Journal*, 35 (3), pp. 442-459.

Schneider, J.A., and Mayne, P.W. (1999) Soil Liquefaction Response in Mid-America Evaluated by Seismic Piezocone Tests, *Report, MAEC Project No. GT-3*, October, pp. 273 (Under Review).

Seed, H.B., and Idriss, I.M. (1971) Simplified procedure for evaluating soil liquefaction potential, *J of Soil Mech and Foundation Div*, ASCE, 97 (SM9), pp. 1249-1273.

Seed, H.B., Idriss, I.M., and Arango, I. (1983) Evaluation of liquefaction potential from field performance data, *J of Geotechnical Eng*, ASCE, 109 (3), pp. 458-482.

Stover, and Coffman (1993) *Seismicity of the United States 1568 - 1989 (Revised)*, U.S. Geological Survey Professional Paper 1527.

Suzuki, Y., Tokimatsu, K., Taye, Y., and Kubota, Y. (1995) Correlation between CPT data and dynamic properties of in-situ frozen samples, *Proc. 3rd Int. Conf. on Recent Adv. in Geotech Earthquake Eng. and Soil Dynamics*, pp. 249-252.

Toro, G.R., Abrahamson, N.A., and Schneider, J.F. (1997) Model of strong ground motions from earthquakes in central and eastern North America: Best estimates and uncertainties, *Seismological Research Letters*, 68 (1), pp. 41-57.

Tuttle, M.P., Lafferty, R.H., and Schweig, E.S. (1998) Dating of Liquefaction Features in the New Madrid Seismic Zone and Implications for Earthquake Hazard, NUREG/GR-0017, U.S. Nuclear Regulatory Commission (in press).

Youd, T.L. (1984) Recurrence of liquefaction at the same site, *Proc. of the 8th World Conf. on Earthquake Eng.*, Vol. III, San Francisco, CA, July 21-28, pp. 231-238.

COMPARISON OF SPT-CPT LIQUEFACTION EVALUATIONS AND CPT INTERPRETATIONS

By Juan I. Baez[1] A.M. ASCE, Geoffrey R. Martin[2] M. ASCE, and T. Leslie Youd[3], M. ASCE

ABSTRACT

The NCEER liquefaction workshop document (Youd and Idriss, 1997) presented the Standard Penetration Test (SPT), and the newer Cone Penetration Test (CPT) methods to undertake liquefaction analyses, and discussed available data that supported the validity of their use. The workshop encouraged verification of CPT soil classification by site specific SPTs. It also stated that the two methods appear to correlate well in terms of predicting the occurrence of liquefaction. However, examination of recent case histories suggest that cyclic resistance ratios (CRR), and in turn factors of safety against liquefaction, may differ depending on the method used; in some cases by as much as ±30%. This paper examines those differences and poses the possibility that in some cases either the CPT or SPT base curves may not be conservative enough, and thus could lead to erroneous conclusions. On the issue of CPT interpretations this paper presents laboratory test results of fines content and Chinese criteria parameters of side-by-side probes for comparison with published CPT correlations.

INTRODUCTION

For many years liquefaction analyses have been performed using the Standard Penetration Test (SPT) and the simplified procedures originally developed by Seed and Idriss (1971, 1982), and updated by Seed et al., (1985) and Youd and Idriss (1997). The procedure essentially compares the cyclic resistance ratio (CRR) against the earthquake and induced cyclic stress ratio (CSR) to determine the possible occurrence of soil liquefaction (or factor of safety). Values of CRR were originally established from

---

[1] Vice President, Hayward Baker Inc., 1780 Lemonwood Drive, Santa Paula, California 93060.
[2] Professor, Department of Civil and Environmental Engineering, University of Southern California, Los Angeles, California.
[3] Professor, Department of Civil and Environmental Engineering, Brigham Young University, Provo, Utah 84602-4081.

17

empirical correlations with the SPT blow count of soil sites which liquefied or did not liquefy during past earthquakes. In the last few years the CPT has gained more acceptance as an alternate, cost-effective, method for measuring penetration resistance and interpreting a description of the soil. Furthermore, a data base from recent earthquake events has been presented by Olson and Stark (1998), and an independent base curve for calculation of CRR from CPT data has been established (Robertson and Wride, 1997). The simplified based curves adopted by the NCEER workshop for both SPT and CPT are reproduced in Figures 1a and 1b.

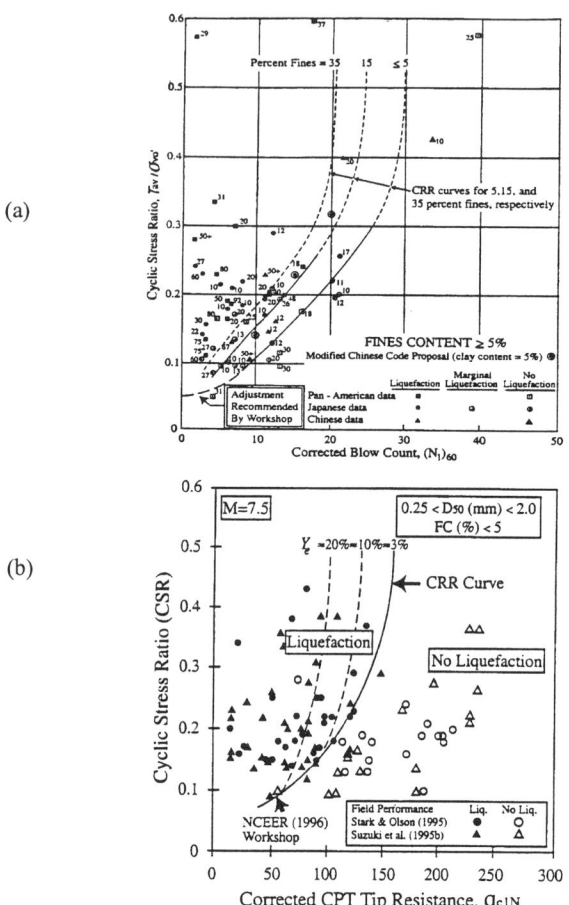

(a)

(b)

Figure 1. Base Curves Recommended by NCEER Workshop (Youd and Idriss, 1997) for Calculation of CRR from (a) SPT and (b) CPT Data

The recommended SPT and CPT base curves for determining the CRR of magnitude 7.5 earthquakes in clean sand can be described by equations as proposed by T. Blake (published in Youd and Idriss, 1997) and Roberston and Wride (1997).

Provided a relationship is known between the SPT blow count and CPT tip resistance for clean sands, it becomes possible to compare the CRR determined independently from the CPT and SPT data sets. While it may not be so critical to determine whether a site has a liquefaction factor of safety of 0.75 or 0.85, for instance, differences may become important around the threshold number which decides whether the site will or will not liquefy, and whether it should be a candidate for remediation. Furthermore, the larger issue may be whether the base curves of the CPT or SPT should be modified because one or the other may be not be conservative enough in predicting the occurrence of liquefaction which may lead to erroneous conclusions.

On the issue of CPT interpretations, the NCEER (1997) liquefaction workshop encourages the geotechnical engineering community to verify the proposed correlations between soil behavior type index, I_c, values and grain size characteristics as well as parameters useful for assessment of the Chinese criteria. Gilstrap and Youd (1998) have examined many case histories on this topic and this paper complements and further discusses that data base.

CPT-SPT CORRELATIONS FOR CLEAN SANDS

Lunne, Robertson, and Powell (1997) reviewed a number of studies of SPT-CPT correlations including, Robertson et al., (1983), Kulhawy and Mayne (1990), and Roberston et al., (1986). These studies were focussed either on the mean particle size, percent fines, or soil behavior type. According to Roberston and Wride (1997), the CRR liquefaction base curve for clean sands (less than 5% fines content) has been derived for soil with mean grain size D_{50} between 0.25 mm and 2 mm. For this range of grain size Robertson et al., (1986) and Kulhawy and Mayne (1990) suggest that a corresponding CPT-SPT, $(q_c/p_a)/N_{60}$ ratio, would be between 4 and 8. Furthermore the Kulhawy and Mayne (1990) study also relates the CPT-SPT to the fines content and suggests that for soils with fines content below 5 percent the $(q_c/p_a)/N_{60}$ ratio ranges between 4 and 7 with an approximate average of about 5. Roberston et al., (1986) suggested $(q_c/p_a)/N_{60}$ ratios for each soil behavior type according to zones, and in the clean sand zone it assigns a ratio of 5.

Jefferies and Davies (1993) and Lunne, Robertson, and Powell (1997) recognized the broad and inaccurate range from previous studies, together with the limitations and variability of the CPT-SPT driving resistance ratio due to the discontinuous nature of the conversion ratio. Further studies led them to conclude the existence of a reasonable relationship between the soil behavior type index, I_c, and the CPT-SPT ratios given by Roberston et al., (1983). The relationship is described with the following equation:

$$(q_c/p_a)/N_{60} = 8.5\ (1 - I_c/4.6)$$ (eq. 1)

In the above equation p_a stands for atmospheric pressure and it is expressed in the same units as the tip resistance, q_c. The I_c value refers to the radius of concentric circles described in the Robertson (1990) classification chart (Figure 2). From Roberston and Fear (1995) and Idriss and Youd (1997), the apparent fines content can be calculated based on the I_c value. In the case of clean sands with less than 5% the corresponding I_c value ranges between 1.3 and 1.55, and therefore based on equation (1) the corresponding $(q_c/p_a)/N_{60}$ ratio ranges between 5.6 and 6.1.

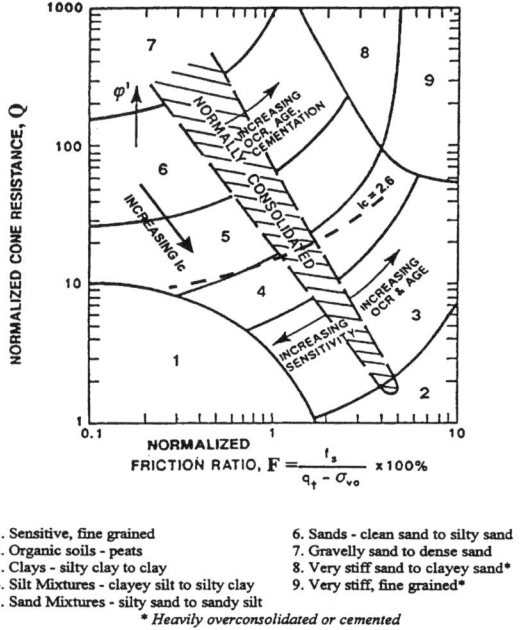

1. Sensitive, fine grained
2. Organic soils - peats
3. Clays - silty clay to clay
4. Silt Mixtures - clayey silt to silty clay
5. Sand Mixtures - silty sand to sandy silt

6. Sands - clean sand to silty sand
7. Gravelly sand to dense sand
8. Very stiff sand to clayey sand*
9. Very stiff, fine grained*

* Heavily overconsolidated or cemented

Figure 2. Normalized CPT Soil Behavior Type Chart (Robertson, 1990)

The authors have tested the validity of equation (1) at several sites where side by side CPTs and high quality SPTs and have been performed. Figure 3 shows a comparison between equation (1) and measured field case histories. For purposes of the comparison the q_c/p_a value has been averaged over a 1 foot continuous interval at the corresponding depth where the SPT was taken.

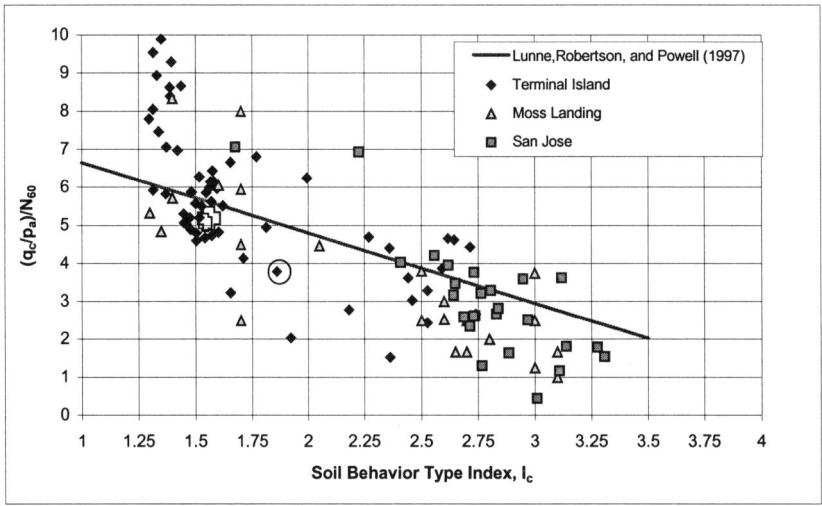

Figure 3. Case Histories versus Theoretical CPT-SPT Correlation with I_c Value

It should be noted that the values of N_{60} have been corrected for rod length, hammer energy, sampler diameter, and liner presence, but not corrected for overburden pressure in order to maintain consistency with previous correlations. It is noted that site specific energy calibrations were not performed at the tested sites, but instead, manufacturer's or published energy values (Seed et al., 1985) were used. Significant scatter is evident in the above data comparison and at first look, it would indicate that agreement with the Lunne, Roberston and Powell (1997) correlation line is poor. However, closer examination reveals that many of the data points may be different than calculated ones because of the location of the SPT in relation to transitional layers, either in type of soil, or penetration resistance of the soil. Thus, a better judgement of the viability of the correlation $(q_c/p_a)/N_{60}$ ratio becomes more evident as the entire depth profile of calculated N_{60} is compared with field measurements of N_{60}. For instance, Figure 4 shows a field case of side-by-side SPT and CPT. The CPT data has been equated to the mentioned correlation N_{60} value and plotted as a continuous profile. Two of the data points, surrounded by an oval and an open cross, can be cross-correlated between Figures 3 and 4. It would seem reasonable to conclude that considering the inaccuracies of the SPT test, the mentioned correlation values provide reasonably appropriate and consistent agreement. A similar conclusion was noted by Martin (1992) in a discussion of energy based correlations between normalized CPT end tip resistance, friction ratio, and $(N_1)_{60}$.

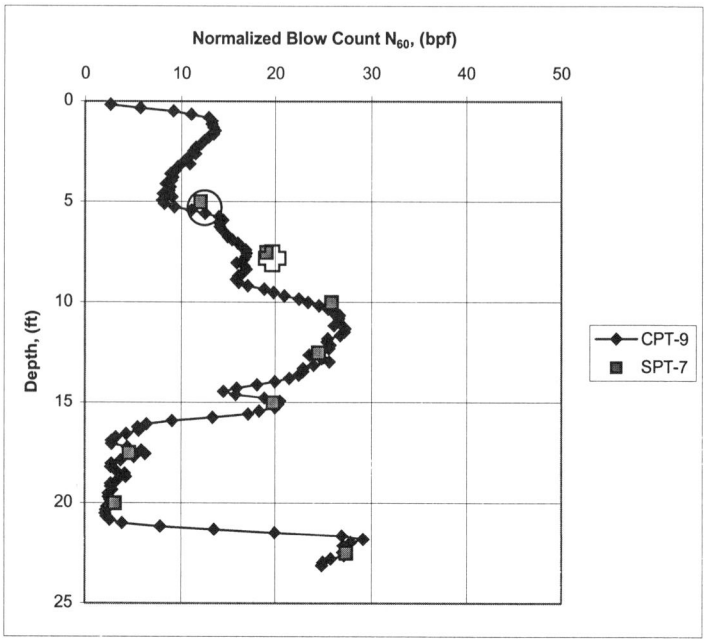

Figure 4. Comparison of Measured SPT N_{60} with Calculated SPT N_{60} for Terminal Island Site

COMPARISON OF SPT AND CPT LIQUEFACTION BASE CURVES

Gilstrap and Youd (1998) examined 159 pairs of CPT and SPT based factors of safety for the occurrence of liquefaction. Figure 5 shows the results of their evaluation. The data shows significant scatter around the perfect correlation line (i.e., 1 to 1). Their regression analysis of the data shows the best fit line which is skewed with respect to the perfect correlation line. The skewed best fit line would suggest that the CPT, on average, could calculate a factor of safety which is greater than the SPT when the factor of safety is less than 1, and smaller when the factor of safety is greater than 1. However, the correlation is rather poor and such conclusion could be highly inaccurate. Clearly, the data would indicate that significant variance can be obtained when comparing the liquefaction analysis by SPT and CPT methods. This, in part, can be explained by the variability of the natural deposits and the SPT test generally being less accurate. Nevertheless, the authors believe that a significant part of the difference could be explained on the basis of the chosen liquefaction base curves for CRR which may inherently provide different results.

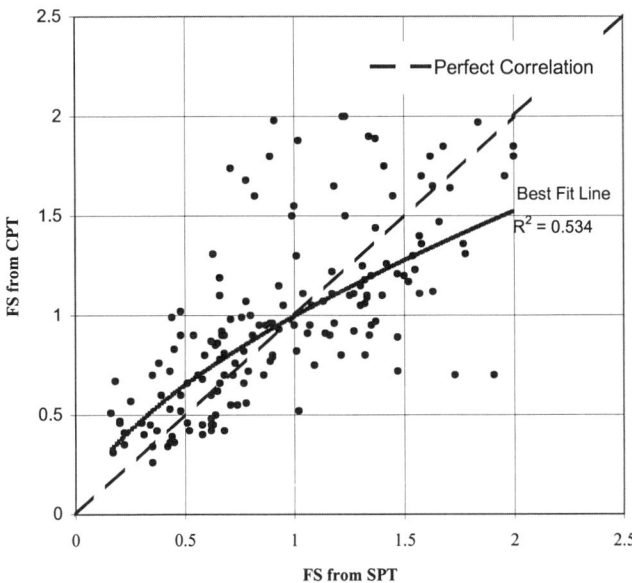

Figure 5. Comparison between Factors of Safety Calculated from CPT and SPT Data Measured at the Case History Sites (Modified after Gilstrap and Youd, 1998)

The preceding section of this paper (CPT-SPT Correlations for Clean Sands), established the viability to reasonably estimate equivalent SPT blow counts from CPT data. It follows that a comparison between independent data sets of SPT and CPT, which were used to construct the liquefaction base curves, is also possible. The comparative study was performed using the SPT-CPT correlation equation (1) from Lunne, Robertson, and Powell (1997), equations that allow the calculation of $CRR_{7.5}$ for clean sands based on corrected blow counts $(N_1)_{60}$ and corrected CPT tip resistance, q_{c1N} of actual case histories.

The results of the study can be evaluated by correlation with different parameters. For instance, the $CRR_{7.5}$ from CPT versus equivalent SPT as derived from equation (1) and shown in Figure 6. The perfect correlation line for CRR (i.e., 1 to 1) is also drawn in this Figure. The data shows a tendency for CRR as calculated from CPT (CRR_{CPT}) data to be lower than equivalent blowcount SPT data below a CRR (CRR_{SPT}) of 0.23, and vice versa above 0.23. In other words, for the same soil and equivalent penetration resistance, the factor of safety against liquefaction is likely to be lower using the CPT when the CRR_{CPT} is below 0.23.

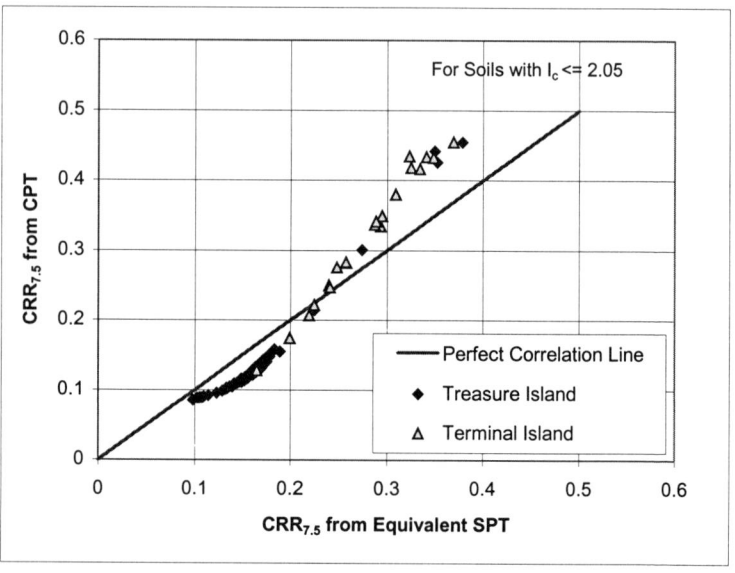

Figure 6. Calculation of $CRR_{7.5}$ from CPT and Equivalent SPT using NCEER (1997) Equations

The $CRR_{7.5}$ threshold value of 0.23 corresponds to a normalized, $(N_1)_{60-cs}$, SPT blow count of 21. The amount of difference in CRR estimates between CPT and SPT over the range of normalized SPT blow counts between 0 and 30 blows per foot can be illustrated as a ratio. This relationship is illustrated in Figure 7. The data suggests that the CPT liquefaction cyclic resistance ratio (CRR_{CPT}) may be up to 25% lower than that calculated by the SPT (CRR_{SPT}) below an $(N_1)_{60-cs}$ value of 21 blows per foot, and up to about 30% higher than the CRR_{SPT} for $(N_1)_{60-cs}$ above a value of 21 blows per foot.

The comparison for CRR from both the CPT and SPT can also be depicted as a superposition on the NCEER (1997) liquefaction base curve for clean sands as shown in Figure 8. The relative conservatism of the CPT for normalized SPT blow counts below 21 becomes evident in this Figure. The Figure also shows the relative SPT conservatism for SPT blow counts over 21.

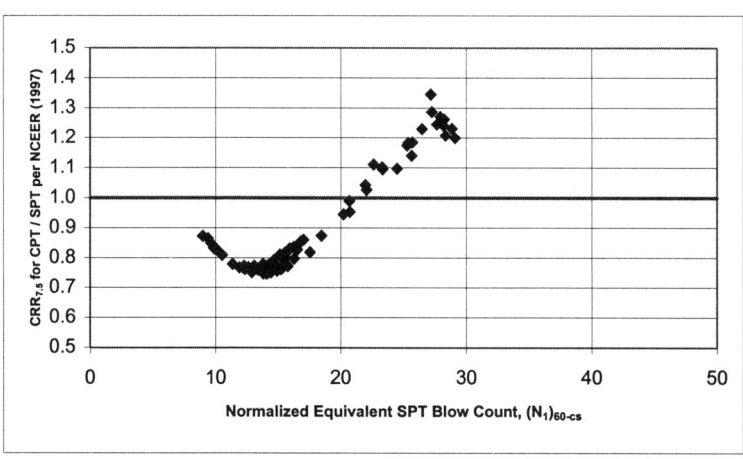

Figure 7. Theoretical CPT/SPT Ratio of CRR$_{7.5}$ with Normalized SPT Blow Count, (N1)$_{60-cs}$

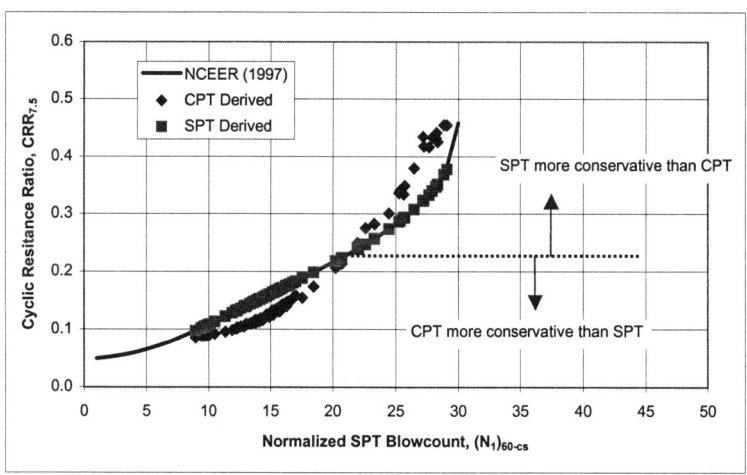

Figure 8. Comparison of CPT CRR$_{7.5}$ with NCEER (1997) Liquefaction Base Curve for SPT in Clean Sands

The difference in liquefaction base curves may be enough to warrant closer examination of the original empirical data for both SPT and CPT and perhaps question whether the base curves need adjustment for better agreement. In the case of the SPT data (Figure 1a) one notes that in the questionable area of discrepancy below for $(N_1)_{60\text{-cs}}$ of 21, there are no empirical data points which did not liquefy for soils with fines content lower than 5%. Conversely, the CPT liquefaction base curve (Figure 1b) shows many data points on the boundary line for cyclic stress ratios (CSR) below 0.23. This would indicate that one could place a higher degree of confidence on the CRR_{CPT} base curve and suggest that perhaps the CRR_{SPT} base curve requires a modification towards lower CRR values for soils with SPT blow counts $(N_1)_{60\text{cs}}$ below 21. Similarly, relatively few points are available from the original empirical SPT data base to accurately judge the appropriateness of the CRR base curve for a CSR greater than 0.23. At this CSR level, more cases are reported in the CRR_{CPT} empirical liquefaction data, including one (q_{c1N} 150, CSR 0.29) which showed liquefaction but would otherwise be considered non-liquefiable according to the placement of the base CRR_{CPT} curve. It is interesting to note that this data point would be considered as marginally liquefiable under the equivalent CRR_{SPT} liquefaction base curve. Despite these observations, CRR base curves should be used with caution as there are not many field cases that have been studied when the CSR exceeds approximate values of 0.4.

CPT CORRELATIONS FOR CHINESE CRITERIA AND FINES CONTENT

Gilstrap and Youd (1998) performed a study to test the validity of Robertson and Wride (1997) opinion indicating that soils with soil type behavior index $I_c > 2.6$ would be too clay rich or too plastic to liquefy. Approximately 6500 data points were compiled by Gilstrap and Youd (1998) to test the I_c boundary for the application of the liquefaction Chinese criteria for clayey soils. The criteria states that soils which meet all of the following characteristics are susceptible to liquefaction:

Clay content (defined as % finer than 0.005 mm)   <15%
Liquid Limit   < 35
Water Content   > 0.9 x Liquid Limit.

Figure 9 shows the results of Gilstrap and Youd (1998) study. They concluded that, indeed, over 90% of the soils with I_c value > 2.6 classified as clayey and met the Chinese criteria as stated above. Furthermore, 25% to 50% of soils with I_c values between 2.4 and 2.6 also met the Chinese criteria, suggesting that in practice soils with I_c value as low as 2.4 could reasonably meet the Chinese criteria, but should be verified by tests on retrieved samples. Approximately 120 sets of laboratory and CPT tests were evaluated by this paper's authors. For comparison, the results of this study are also plotted in Figure 9. Although not as many tests were available for evaluation, the findings are similar to those of Gilstrap and Youd (1998) except that a higher percentage of soils with I_c values between 2.4 and 2.6 may be classified as nonliquefiable per the Chinese criteria.

Andrews and Martin (2000) have also studied the Chinese criteria and confirmed its applicability, but recommend that clay contents (defined as grains finer than 0.002 mm rather than 0.005 mm), greater than 10% would render the soils nonliquefiable if, in addition, the liquid limit is greater than, or equal to 32. Andrews and Martin (2000) note that based on work by Koester (1992), a liquid limit of 35 determined by the Chinese criteria (using the People's Republic of China fall cone penetrometer), is equivalent to a liquid limit of about 32 as determined by the Casagrande-type percussion device.

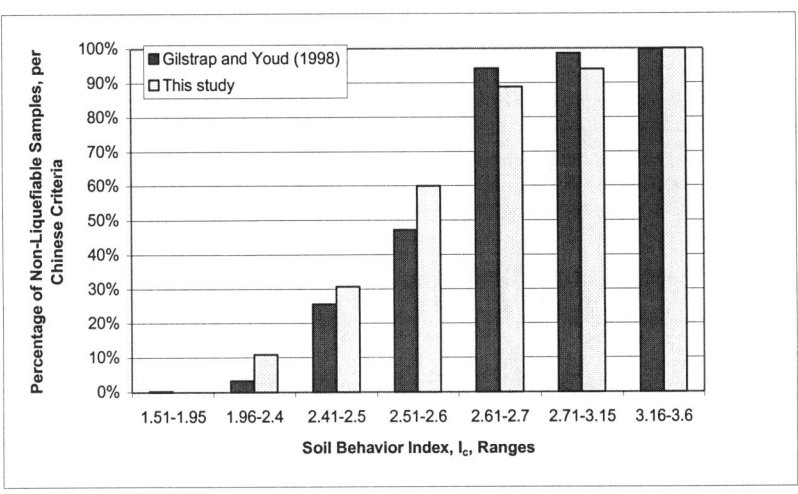

Figure 9. Percentage of Samples Meeting the Chinese Criteria (no liquefaction) for Given I_c Value

One of the shortcomings of the CPT has been its inability to routinely obtain samples for visual and laboratory testing. Attempts have been made by various researchers including Kulhawy and Mayne (1990), Roberston and Wride (1997), and others to correlate CPT values and approximate fines content. Although the I_c value depends on factors such as mineralogy, sensitivity, and stress history, Roberston and Wride (1997) found that an approximate correlation between I_c and apparent fines content is possible. Figure 10 shows their recommended general correlation.

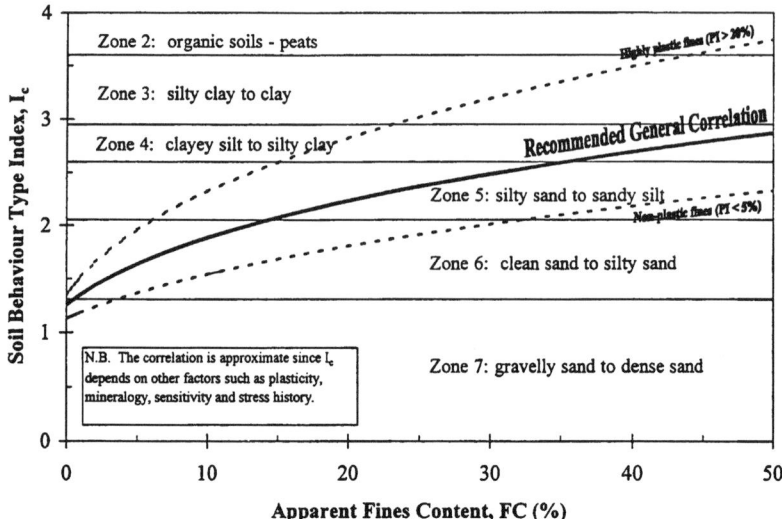

Figure 10. Variation of CPT Soil Behavior Index Value, I_c, with Apparent Fines Content (Roberston and Wride, 1997)

As many practicing engineers often use the Robertson and Wride (1997) recommended correlation to estimate fines content, the authors decided to evaluate it against case histories. Gilstrap and Youd (1998) compared 184 pairs of measured fines content and I_c values against Robertson and Wride (1997) recommended correlation. They found that for I_c values greater than 2.6, the recommended correlation under predicted the fines content for approximately 90% of the data.

The Gilstrap and Youd (1998) data base together with other case histories (5 sites) from this study have been plotted in Figure 11. A regression analysis was performed using various function types. The best fit was obtained with a power function as shown in Figure 11. Although the statistical fit is poor to marginal ($R^2 = 0.62$), the best fit relationship appears to show improvement over the recommended Robertson and Wride (1997) correlation for the sites of concern. It is noted that this correlation and the "best fit" correlation provide similar results up to an I_c value of about 2.05 (approximate fines content of 14%) and consequently the Roberston and Wride (1997) correlation is validated to this point. However, the data for the case histories evaluated suggests that Robertson and Wride (1997) general correlation would have, in most cases, overpredicted the amount of fines content when the I_c values is higher than 2.05. Although it is recognized that the data evaluated in this study may be more limited in comparison with the Roberston and Wride (1997) data, it suggests that a revised correlation formula that may be appropriate is as follows:

If $I_c < 1.26$ Apparent fines content, FC (%) = 0   (2a)

If $1.26 <= I_c <= 2.05$ Apparent fines content, FC (%) = $1.75 \, I_c^{3.25} - 3.7$ (2b)

If $2.05 < I_c <= 2.82$ Apparent fines content, FC (%) = $(0.75 \times I_c)^{6.14}$ (2c)

If $I_c > 2.82$ Apparent fines content, FC (%) = 100   (2d)

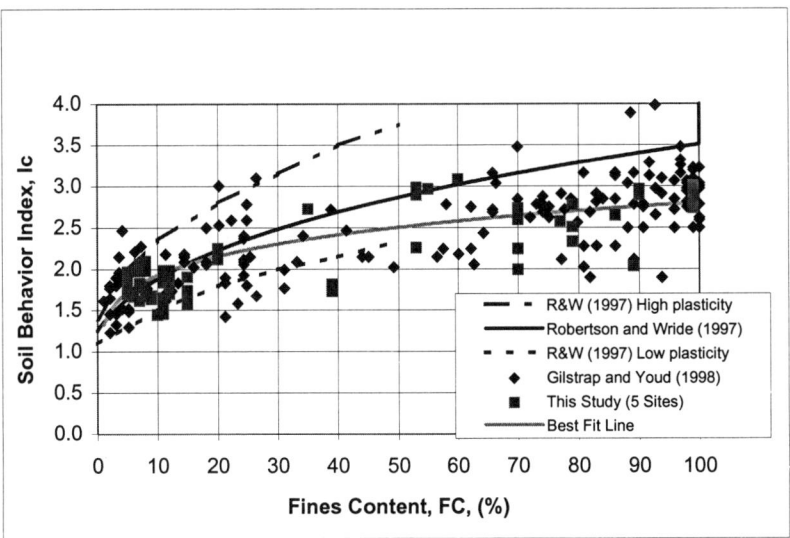

Figure 11. Case Histories Measured Fines Content Versus Soil Behavior Index, I_c

While CPT based correlations to estimate fines content may be convenient, the authors strongly suggest that they not be used solely, or in lieu of prudent collection of samples and laboratory testing. Due to high variability of soil mineralogy, stress history, and plasticity characteristics, among others, reliable site specific correlations may be difficult to achieve.

CONCLUSIONS

Researched case histories of side by side CPT and SPT data indicates that the correlation for penetration resistance versus I_c value as presented by Lunne, Roberston and Powell (1997) provides reasonable agreement. Observed scatter may be largely attributed to the inconsistency of the SPT test and the numerous correction factors, as well as presence of transitional materials around the location of the SPT. The use of a validated SPT-CPT correlation of penetration resistance allowed a comparison of CRR value as calculated from the CPT and from an equivalent SPT value recommended in the NCEER liquefaction workshop (Youd and Idriss, 1997). The case histories researched indicated that for the same soil and equivalent penetration resistance, the factor of safety against liquefaction is likely to be lower using the CPT when the CRR is below 0.23 and vice versa above 0.23. The $CRR_{7.5}$ threshold value of 0.23 corresponds to a normalized SPT blow count of 21. The data suggests that the CPT liquefaction cyclic resistance ratio (CRR) may be up to 25% lower than that calculated by the SPT below an $(N_1)_{60-cs}$ value of 21, and up to about 30% higher than the CRR calculated by the SPT for $(N_1)_{60-cs}$ above a value of 21 blows per foot.

The differences between CPT and SPT derived cyclic resistance ratio (CRR), and in turn the liquefaction factor of safety, can, in some cases, be critical for deciding whether the soil is likely to liquefy or not. Although the case histories investigated showed significant discrepancy between CRR_{CPT} and CRR_{SPT}, more data and analyses is needed before concluding that the recommended liquefaction base curves should be modified to achieve better agreement. It is noted, however, that the data base for the CRR_{CPT} curve appears, to date, more extensive than that of the CRR_{SPT} base curve and proper engineering judgement should be exercised in the choice of base curves for liquefaction analyses.

Evaluations for CPT derived correlations against the liquefaction Chinese criteria of the case histories evaluated suggests that soils with I_c values as low as 2.4 could reasonably meet the criteria. Extensive research by Gilstrap and Youd (1998) showed that for their data base, and additional case histories from this study, the Chinese criteria was met in more than 90% of soils with an I_c value of 2.6 or greater, and thus companion sampling and laboratory index testing should accompany CPT liquefaction investigations.

With regard to CPT derived correlations for predicting apparent fines content, evaluated case histories support the findings of Gilstrap and Youd (1998) which indicate that Roberston and Wride (1997)'s recommended general correlation tends to overpredict the fines content for high Ic values. Nevertheless, the data appears to fall largely within the wide error bar of Robertson and Wride (1997). While CPT correlations to estimate fines content may be convenient, the authors strongly suggest that they not be used solely, or in lieu of prudent collection of samples and laboratory testing. Due to high variability of soil mineralogy, stress history, and plasticity characteristics among others, site specific correlations may be difficult to achieve.

ACKNOWLEDGMENTS

The authors gracefully acknowledge the many contributions of geotechnical consultants from Geosoils Consultants, Van Nuys, CA, Leighton and Associates, Irvine, CA, Diaz & Yourman, Tustin, CA, Dames & Moore, San Jose, CA, Fugro West, Ventura, CA, NMG Geotechnical, Irvine, CA, and many others who provided valuable case history data for analysis.

REFERENCES

Andrews D.C.A., and G.R. Martin (2000), "Criteria for Liquefaction of Silty Soils," 12th World Conference on Earthquake Engineering, Auckland, New Zealand.

Blake, T.F. (1996) Personal communication with T.L. Youd, 1996.

Gilstrap, S.D., and T.L. Youd (1998), "CPT Based Liquefaction Resistance Analyses Evaulated Using Case Histories," Master of Science Thesis, Technical Report CEG-98-01, Department of Civil and Evironmental Engineering, Brigham Young University, Provo, Utah.

Jeffries, M.G., and M.P. Davies (1991), "Use of CPTu to Estimate Equivalent SPT N60," ASTM Geotechnical Testing Journal, 16(4): 458-467.

Koester, J.P. (1992), "The Influenec of Test Procedure on Correlation of Atterberg Limits with Liquefaction in Fine – Grained Soils," Geotechnical Testing Journal, Vol. 15, No 4, pp 352-360.

Kulhawy, F.H., and P.H. Mayne (1990), "Manual on Estimating Soil Properties for Foundation Design," Electric Power Research Institute, EPRI, August 1990.

Lunne, T., P.K. Roberston, and J.J.M. Powell (1997) Cone Penetration Testing in Geotechnical Practice, 1st edition, Blackie, Academic and Professional, London, UK 312 p.

Martin, G.R. (1992), "Soil Property Evaluation for Earthquake Stability and Deformation Analyses," Invited state-of-the art paper at the ASCE Specialty Conference on "Stability and Performance of Slopes and Embankments II," University of California, Berkeley, June, 1992.

Martin, G.R., and M. Lew, (1999), "Recommended Procedures for Implementation of DMG Special Publication 117, Guidelines for Analyzing and Mitigating Liquefaction in California," Southern California Earthquake Center.

Olson, S.M. (1997), "Cyclic Liquefaction Based on the Cone Penetrometer Test," Proceedings of the NCEER Workshop on Evaluation of Liquefaction Resistance of Soils, Youd T.L. and Idriss I.M., editors, Technical Report NCEER-97-0022.

Olson, S.M., and T.D. Stark (1998), "CPT Based Liquefaction Resistance of Sandy Soils," Geotechnical Earthquake Engineering and Soil Dynamics III, ASCE GSP 75, pp. 325-336.

Robertson, P.K., R.G. Campanella, and A. Whightman (1983), "SPT-CPT Correlations," Journal of Geotechnical Engineering, ASCE, 109 (11), 1449-59.

Robertson, P.K., R.G. Campanella, D. Gillespie, and J. Creig (1986), "Use of Piezometer Cone Data," Proceedings of the ASCE Specialty Conference In Situ '86, 1263-80.

Robertson, P.K. and C.E. Fear (1995), "Liquefaction of Sands and its Evaluation. IS TOKYO '95," First International Conference on Earthquake Geotechnical Engineering, Keynote Lecture, November 1995.

Robertson, P.K. and C.E. (Fear) Wride (1997), "Cyclic Liquefaction and its Evaluation Based on the SPT and CPT," Proceedings of the NCEER Workshop on Evaluation of Liquefaction Resistance of Soils, Technical Report NCEER-97-0022.

Seed H.B., and I.M. Idriss (1971), "Simplified Procedure for Evaluating Soil Liquefaction Potential," *Journal of Soil Mechanics and Foundations Division*, ASCE, Vol. 97.

Seed H.B., and I.M. Idriss (1982), "Ground Motions and Soil Liquefaction During Erathquakes," Earthquake Engineering Research Institute Monograph.

Seed H.B., K. Tokimatsu, L.F. Harder, and R.M. Chung (1985), "Influence of SPT Procedures in Soil Liquefaction Resistance Evaluations," *Journal of Geotechnical Engineering*, ASCE, Vol. 111, No 12, pp. 1425-1445.

Stark, T.D. and S.M. Olson (1995), "Liquefaction Resistance Using CPT and Field Case Histories," *Journal of Geotechnical Engineering*, ASCE, Vol. 121, No. 12, pp. 856-869.

Susuki, Y., K. Tokimatsu, K. Koyamada, Y. Taya, and Y. Kubota (1995), "Field Correlation of Soil Liquefaction Based on CPT Data," Proceedings of the International Symposium on Cone Penetration Testing, CPT'95, Linkoping, Sweden, Vol. 2, pp. 583-588.

Tokimatsu, K. and H.B. Seed (1987) "Evaluation of Settlement in Sands due to Earthquake Shaking," Journal of the Geotechnical Engineering Division, ASCE, Vol. 113, No 8, August.

Youd, T.L. and I.M. Idriss (1997), "Proceedings of the NCEER Workshop on Evaluation of Liquefaction Resistance of Soils," Technical Report NCEER-97-0022.

IN SITU INVESTIGATION OF LIQUEFIED GRAVELS AT SEWARD, ALASKA

Joseph P. Koester[1], Chris Daniel[2], and Michael Anderson[3]

ABSTRACT

Dynamic in situ penetration tests and crosshole shear wave velocity measurements were performed in deep alluvial gravel deposits at Seward, Alaska that were shaken and reportedly liquefied by the March 27, 1964 Alaska Earthquake. Two sizes each of split spoon and dynamic cone penetrometers were driven into the gravels near the mouth of the Resurrection River that had exhibited settlement and lateral spreading consequent to earthquake shaking. Two safety hammers of different size were employed, and the energy delivered using various hammer and penetrometer combinations was measured throughout all tests, with the exception of two duplicate tests of the larger sampler and dynamic cone. Limited measurements were also made of hammer velocity, using a radar system developed for that purpose, to allow for kinetic energy determination. Disturbed soil samples were evaluated for classification and to determine stratigraphy. This paper summarizes the procedures used at the Seward site, as well as preliminary correlation of penetration resistance, measured by alternate means and normalized for overburden stress and driving energy, and shear wave velocity in the coarse-grained deposits. Comparisons are also made to penetration resistance measurements made by the Alaska Highway Department immediately following the 1964 earthquake.

[1] Research Civil Engineer, CEERD-GG-E, U.S. Army Engineer Waterways Experiment Station, 3909 Halls Ferry Road, Vicksburg, Mississippi 39180-6199

[2] Graduate Student, In-Situ Testing Group, Civil Engineering Department, University of British Columbia, 2324 Main Mall, Vancouver, BC, Canada V6T 1Z4

[3] Civil Engineer, CEPOA-EN-ES-SG, U.S. Army Engineer District, Alaska, P.O. Box 898, Anchorage, Alaska 99506-0898

INTRODUCTION

In situ material properties of soils containing gravels and cobbles, including inference of liquefaction resistance and residual strength, are currently estimated by converting results of any of several dynamic penetration tests (Becker Hammer Penetration Test - BPT, Large Penetration Test - LPT, and site-specific variations) to equivalent Standard Penetration Test (SPT) blowcounts. BPT/SPT correlations employed to date for U. S. Army Corps of Engineers projects have considered diesel combustion efficiency (in the manner described by Harder and Seed, 1986 and Harder, 1988). Field measurements using Pile Driving Analyzer (PDA) equipment have also shown that BPT efficiency is strongly influenced by mechanical energy losses, including friction acting along the driven casing (Sy, 1993, Sy and Campanella, 1993 and Sy and Campanella, 1994).

The U. S. Army Engineer Research and Development Center (ERDC), Waterways Experiment Station (WES) is investigating in situ testing methods applied to determine the response of various soil types to earthquake ground motions, as part of a research program on seismic stability evaluation of reservoir dams. A limited program of dynamic in situ penetration tests was performed in deep alluvial gravel deposits at Seward, Alaska that were shaken and apparently liquefied by the March 27, 1964 Alaska Earthquake. Two sizes of split spoon samplers and dynamic cone penetrometers of two sizes were driven into the gravels near the mouth of the Resurrection River that had exhibited settlement and lateral spreading consequent to earthquake shaking. It was the primary author's original intention to perform BPT's in the Seward gravels, but mobilization costs for a suitable BPT rig were prohibitively high.

Shear wave velocities were measured at the site in September 1999 to a depth of 75 ft by crosshole and downhole procedures, using a pair of 4-inch (10 cm) inside diameter PVC-cased borings installed by rotary drilling and spaced 10.4 ft (3.2 m) apart. This paper describes the preliminary interpretation of the geophysical data obtained at the site.

BACKGROUND

Liquefaction and consequent post-earthquake instability has been observed during moderate to strong earthquakes in loose to medium dense gravels, both naturally deposited and placed by various artificial processes. Table 1 lists field occurrence data on liquefaction in gravels during earthquakes worldwide during the last 20 years. In most cases, liquefaction occurrence is actually only confirmable through surface expression, such as sand boils or lateral spreads where there are no structures loading the ground, or by subsurface instrumentation that registers effective stress state or dynamic response. In situ tests are not yet developed to confirm or deny the occurrence of liquefaction in any soils, and gravels are difficult to test in situ. The BPT has been widely used of late to sample and characterize stiffness of coarse grained deposits, and shows promise as a tool

Table 1. **Observations of Liquefaction in Gravelly Soils**
(adapted from personal communication, Alex Sy, Klohn-Crippen Consultants, Ltd,
Richmond, BC, Canada, 1996)

YEAR	M	EARTHQUAKE	REFERENCE
1981	7.9	Mino-Owari, Japan	Tokimatsu and Yoshimi (1983)
1943	7.3	Fukui, Japan	Ishihara (1985)
1964	9.2	Valdez, Alaska	Coulter and Migliaccio (1966)
1975	7.3	Haicheng, PRC	Wang (1984)
1976	7.8	Tangshan, PRC	Wang (1984)
1978	7.4	Miyagiken-Oki, Japan	Tokimatsu and Yoshimi (1983)
1983	7.3	Borah Peak, Idaho	Youd, et al. (1985), Harder (1988)
1988	6.8	Armenia	Yegian, et al. (1994)
1992	5.8	Roermond, Netherlands	Maurenbrecher, et al. (1995)
1993	7.8	Hokkaido, Japan	Kokusho, et al. (1995)
1995	6.8	Kobe, Japan	Kokusho, et al. (1995)

to investigate properties associated with liquefaction potential evaluation. Density tests are extremely valuable, but are usually impractical to conduct in saturated gravel deposits at depths greater than about 10 ft (3.05 m). Sampling and testing of gravels is difficult, but in situ freezing techniques such as those employed in Japan (e.g., Tanaka, et al., 1989) have been successfully used for recovery of high-quality specimens of coarse-grained soils.

OBJECTIVES

The objectives of the experiments discussed in this paper were to apply and compare results from a variety of in situ tests in a gravelly deposit that exhibited behavior indicative of earthquake-induced liquefaction. A companion laboratory chamber investigation was also performed to evaluate particle size and confining pressure effects on penetration resistance and seismic velocities, and a report is being developed for later publication. The overall goals of these studies are to improve techniques to estimate in situ cyclic and residual strengths of soils empirically and evaluate alternate in situ testing technologies for measuring in situ parameters (such as density or void ratio) needed to characterize dynamic strength.

PREVIOUS INVESTIGATIONS

The experiment site was selected based on reports of damage to the bridges along the Alaska Railroad and those along the Anchorage-to-Seward highway that resulted from liquefaction and lateral spreading of the gravelly soils beneath and along the streambanks at each abutment, caused by the March 27, 1964 Alaska Earthquake (McCulloch and Bonilla, 1970). Bridge piers at each crossing of the braided stream channel of the Resurrection River at Seward were founded on piles driven to roughly 40 ft (12.2 m) depth. As a result of 1964 earthquake shaking, these piles were displaced streamward without rotation, suggesting deep-seated spreading such as might accompany liquefaction. The Alaska State Highway Department performed a limited series of dynamic penetrometer and split spoon penetration tests at the site following the 1964 earthquake, but no other in situ tests have been conducted there until the experiments discussed herein.

The previous dynamic penetrometer tests, reported by McCulloch and Bonilla (1970), were conducted by continuously driving a 2.5-inch (0.064 m) diameter solid cone (essentially a plugged split- spoon penetrometer of the same size) with a 342-lb (1521 N) safety hammer (the actual sliding weight, not including anvil and driving rod is 294 lbs, or 1308 N). The safety hammer was dropped 30 inches (0.762 m), and blows per foot were recorded and published in their report as N-values. A limited number of Standard Penetration Tests were also conducted at that time using a 140-lb (623 N) safety hammer and a 2-inch (0.051 m) diameter sample spoon with a 30-inch (0.762 m) hammer drop (Figure 1). Those tests indicated loose, gravelly sands and sandy gravels to significant (greater than 75 ft, or 22.9 m) depth. Depth to bedrock in this filled fjord of the Exit Glacier is not known. The soils are mainly normally consolidated alluvial outwash from one or more upland glaciers in the area.

FIELD PROCEDURE SUMMARY

The field exercises described in this paper took place at Seward, Alaska, during the periods 28 August - 5 September 1998 and 27-30 September 1999. Figure 2 is an aerial photo of the Resurrection River - Mineral Creek floodplain at Seward, showing test locations for this project. The site is on property owned by the Metco Construction Materials Company, a gravel and aggregate harvesting operation at Seward. A drill crew from the U. S. Army Engineer District, Alaska (CEPOA), deployed and operated a tracked drill rig and the various safety hammer and rod assemblies used in this study. The hammers, samplers, and drive points used in the present investigation are shown in Figure 3.

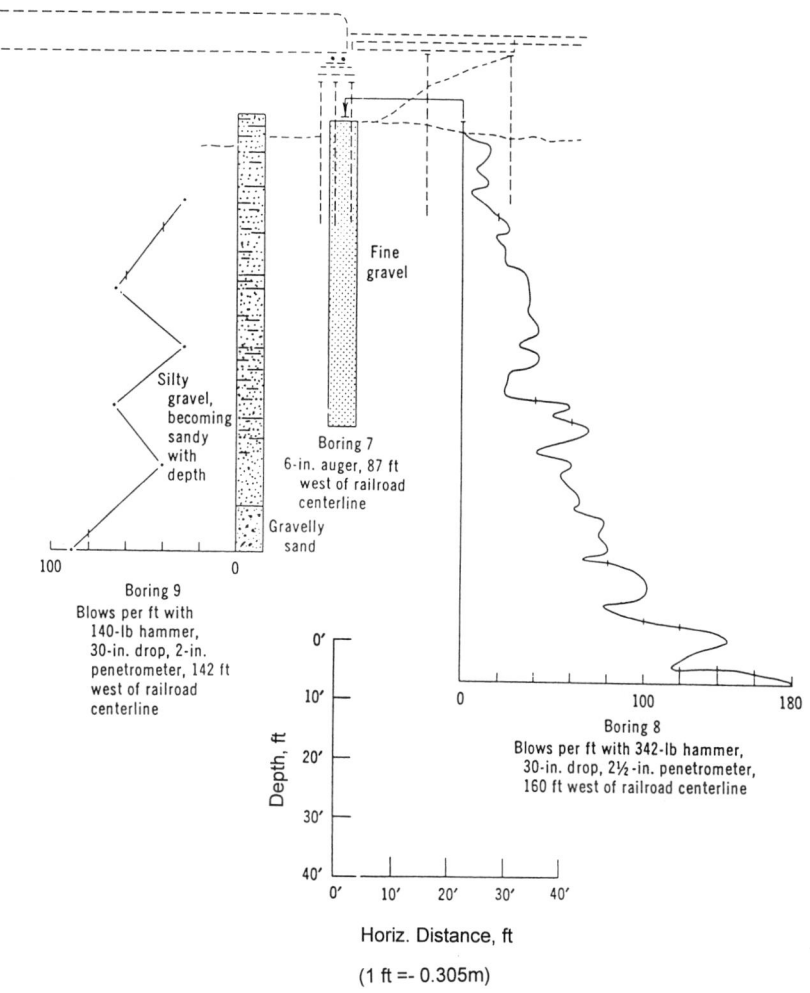

Fine
gravel

Silty
gravel,
becoming
sandy
with
depth

Boring 7
6-in. auger, 87 ft
west of railroad
centerline

Gravelly
sand

100   0

Boring 9
Blows per ft with
140-lb hammer,
30-in. drop, 2-in.
penetrometer, 142 ft
west of railroad
centerline

0'

10'

Depth, ft

20'

30'

40'

0'   10'   20'   30'   40'

0   100   180

Boring 8
Blows per ft with 342-lb hammer,
30-in. drop, 2½-in. penetrometer,
160 ft west of railroad centerline

Horiz. Distance, ft

(1 ft =- 0.305m)

**Figure 1. Borings and Penetration Tests Conducted at Seward Highway Bridge
at Mile 3.3 During May 1964 (Two Months after the March 27, 1964
Great Alaska Earthquake) (after McCulloch and Bonilla, 1970)**

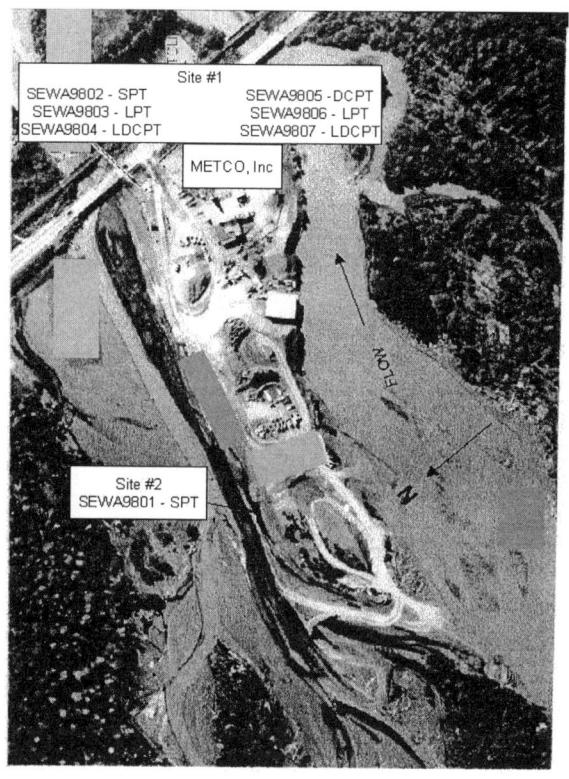

Figure 2. Aerial view of Seward, Alaska, Test Site

**Figure 3. Safety Hammers, Split Spoon Samplers, and Solid Drive Tips
Used in This Study.**
From top to bottom, central: 342 lb hammer; 2.5 in. I.D. sampler; 140 lb hammer;
1.5 in. I. D. sampler. Also shown, center right, is the 3.5 in. diameter LDCPT tip;
the 2.5 in. diameter DCPT tip is at lower right (1 lb = 4.45 N; 1 in. = 0.0254 m)

The first penetration test was conducted in the riverbed itself, in a sand-and-gravel bar
about 400 ft (122 m) upstream from the bridge at Seward Highway Mile 3.3. This first
hole was sampled continuously using a standard split spoon (2-inch, or 0.051 m O.D., 1.4-
inch, or 0.036 m I.D.) sampler and was advanced using a 4-inch (0.102 m) I.D. hollow stem
auger (HSA). Driving energy was measured and recorded for every hammer blow, using
a dynamic energy measurement transducer (Figure 4) on loan from the British Columbia
Hydropower Authority (BC Hydro). The transducer consists of a set of strain gages and
accelerometers mounted near the center of a section of drill rod; these in turn allow for
measurement of dynamic force and velocity. Borehole water level was maintained near the
ground surface during all drilling operations, especially during removal of the sand plug,
or pilot bit. Hole depths were always measured after pilot bit removal, to determine
whether any heave had occurred; heave at the site was generally less than 0.1 ft (0.03 m)
using this procedure. The first hole was drilled to 60 ft (18.3 m) below ground surface;

Figure 4. Dynamic Energy Measurement Transducer
(loaned by British Columbia Hydropower Authority)

most sampling drives retained material (predictably low retention with the small, standard spoon); all sampled soil was shipped to WES for gradation and other index testing.

The first penetration test of this series was conducted to verify proper operation of all energy measurement gear and to establish careful drilling procedures (including conscientious observance to tighten every rod connection and adapter fitting within the rod string using two wrenches, to minimize stress wave reflections in the rod string). Following the first test, the drill rig was relocated to an abutment area of the highway bridge at mile 3.3. Subsequent overnight heavy rains caused the braided stream to rise four feet and remain at a high water stage for two days. Such rapid variation of flow is typical at this site, and indicates the depositional energy associated with the gravel bed placement. All remaining penetration tests were conducted in the vicinity of the bridge abutment. Table 2 is a summary of investigations conducted during the 1998 experiment series.

A pair of boreholes were installed at the abutment site during August 1999 to support crosshole and downhole shear wave velocity surveys. The holes were rotary drilled by a contractor to CEPOA and PVC casing was installed and grouted following ASTM Standard Method D4428 (ASTM, 1991). Borehole deviation and seismic velocity measurements were made in September 1999; preliminary interpretations of velocity data acquired at that time are discussed below. Directional hammer blows (sledgehammer against a secured, transverse timber for downhole surveys, and a clampable downhole sliding hammer for crosshole surveys) were used as the seismic energy source in all cases.

Table 2. Field Test Summary, Seward Site

Test name /Location	Test Type*	Hammer Weight, lb. (N)/Drop Height, in. (m)	Maximum Depth, ft (m)	Test Spacing
SEWA9801 / Site 2	1.52 in. (0.039 m) ID Split Spoon (SPT)	137 (609 N)/30 (0.762 m)	59.57 (18.16 m)	Continuous to 30 ft (9.14 m), followed by 5 ft (1.52 m) spacing
SEWA9802 / Site 1	1.52 in. (0.039 m) ID Split Spoon (SPT)	137 (609 N)/30 (0.762 m)	79.15 (24.12 m)	5 ft (1.52 m) starting at 15 ft (4.57 m) depth
SEWA9803 / Site 1	2.52 in. (0.064 m) ID Split Spoon (LPT)	294 (1308 N)/34 (0.864 m)	78.9 (24.05 m)	5 ft (1.52 m) starting at 15 ft (4.57 m) depth
SEWA9804 / Site 1	3.5 in. (0.089 m) OD Dynamic Cone (LDCPT)	294 (1308 N)/34 (0.864 m)	43 (13.11 m)	Continuous starting at 15 ft (4.57 m) depth
SEWA9805 / Site 1	2.32 in. (0.059 m) OD Dynamic Cone (DCPT)	294 (1308 N)/30 (0.762 m)	50 (15.24 m)	Continuous starting at 15 ft (4.57 m) depth
SEWA9806 / Site 1	2.52 in. (0.064 m) ID Split Spoon (LPT)	294 (1308 N)/34 (0.864 m)	81 (24.69 m)	5 ft (1.52 m) starting at 15 ft (4.57 m) depth
SEWA9807 / Site 1	3.5 in. (0.089 m) OD Dynamic Cone (LDCPT)	294 (1308 N)/34 (0.864 m)	39.5 (12.04 m)	Continuous starting at 15 ft (4.57 m) depth

* ID = Inner Diameter, OD = Outer Diameter

TEST RESULTS

Penetration test results for all variations of procedure used at the Seward site are shown in Figure 5, uncorrected for overburden pressure, differences in equipment, or procedure. In all cases, three turns of rope were wrapped around the drill rig cathead winch; there was no trip mechanism used, and drop heights and hammer weights are noted in Table 2. The first test (SEWA9801) was performed in the river bed (termed as Site 2 in the figure), and the depths of each of its blowcounts are shown offset by 15 ft (4.57 m) to account for the fact that the surface elevation is lower here than at the bridge abutment, where the remainder of tests were performed. The tests were performed along an upstream-to-downstream line, to avoid lateral variations in streambed deposition as much as possible. It appears from Figure 5 that this strategy paid off; there is remarkable correspondence between penetration

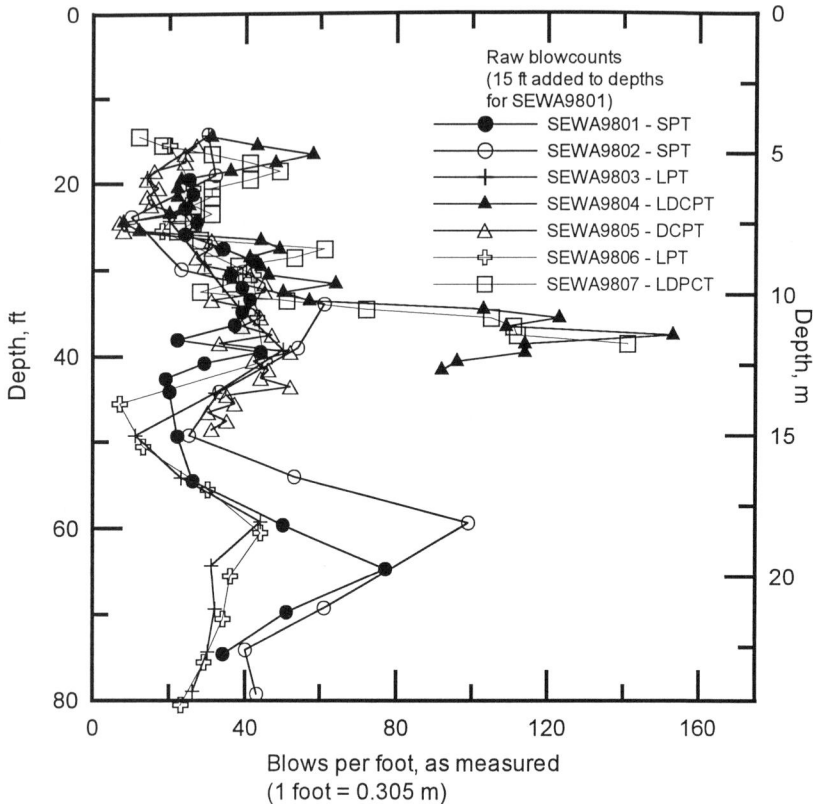

**Figure 5. Penetration Resistance at Seward, As Measured During This Study
(See Table 2 for Legend Explanation)**

resistance, even uncorrected, measured at the two sites. Abbreviations used in the plot are explained in Table 2.

The raw data plot of Figure 5 demonstrates that penetration resistance is strongly dependent on probe size in the gravelly sand and sandy gravels at the Seward site. Penetration resistance was higher among the drive sampler tests when the smaller diameter SPT was employed, but higher penetration resistance was recorded for the larger of the two driven solid cones. Larger particles appear to govern penetration resistance for intruding probes when no sample is obtained, in effect creating an increased footprint for a given driving energy input. The reverse is true, however, with drive samplers, since larger particles are captured within larger cutting shoe areas and impede penetration to a lesser degree. Large particles increase the footprint of the cutting shoe if they will not fit inside. The zone of relatively low blowcounts observed in the Site 1 LPT and SPT, and Site 2 SPT was also encountered in the penetration testing performed following the 1964 earthquake (see Figure 1).

Fines contents ranged generally from 4 to 15 percent by weight in all but four of fifty-one sieve analyses of samples recovered in split spoon samplers at the Seward site. Two samples recovered using a 1.4-inch (0.036 m) I.D. sampler (in the boring designated SEWA9802 in Figure 5) consisted of non-plastic silty sands with: 37 percent fines and 48 percent fines, at depths of about 25 ft (7.62 m) and 50 ft (15.24 m), respectively. Two samples recovered using a 2.5-inch (0.064 m) I.D. sampler (in the boring designated SEWA9803 in Figure 5) were found to have high fines contents: 80 percent and 68 percent, at depths of 45 ft (13.7 m) and 50 ft (15.2 m), respectively. These were both found to have liquid limits of 45, with plasticity indexes of 15 and 11, respectively. The high fines content samples were found at depths where blowcounts were low relative to the majority of data in all three split spoon borings.

To permit comparison of the penetration resistance data in these experiments to published correlations, blowcounts were adjusted to account for the influences of effective overburden pressure and driving energy wherever possible. Energy delivered by the safety hammers to the top of the drill rod string was dynamically measured for all but two duplicate tests in this study, using the BC Hydro transducer system. An "energy ratio," ER, was developed based on data acquired with the system, defined as the percentage of delivered energy relative to the theoretical maximum energy delivered by free-fall of the specific safety hammer used when dropped the distance specific to each test. As a preliminary analysis, the overburden pressure-corrected blowcounts in each test were multiplied by the equipment-specific energy ratio and divided by 60 to "correct" driving energy to a common reference. Both adjustments were accomplished by application of the following expression:

$$(N_1)_{60} = (N \cdot C_N) (ER / 60) \tag{1}$$

where N is the raw blowcount as recorded per foot of penetration in a specific test; and C_N is an adjustment factor to account for effective confining stresses other than 1 tsf (about 100 kPa), determined from the following (NCEER, 1997):

$$C_N = (P_a / \sigma'_v)^{0.5} \tag{2}$$

where P_a is atmospheric pressure, and σ'_v is effective overburden pressure at depth (both must be the same units). Numerous corrections or adjustments have been developed to account for variations among SPT data alone, given variability in procedures and equipment employed for that one, so-called "standard" test (e.g., Youd and Idriss, 1997). Figure 6 is intended only to indicate the consequence of applying the energy adjustment for the equipment used in this study; it is by no means a rigorous consideration of all variables,

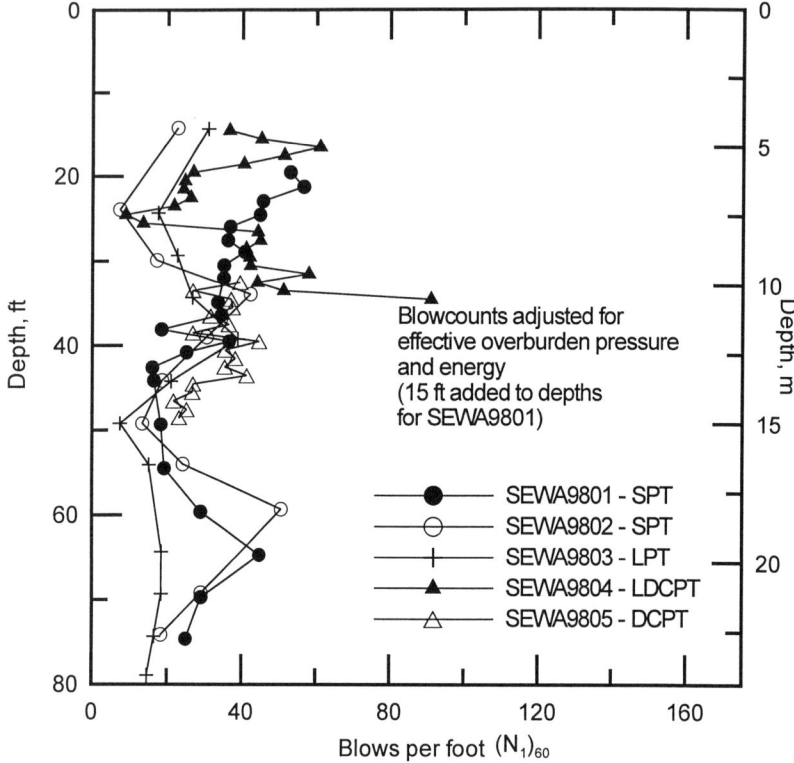

Figure 6. **Penetration Resistance at Seward, Corrected for Overburden Pressure and Driving Energy (See Table 2 for Legend Explanation)**

nor is it intended to allow for future comparison with results from tests where other equipment may be used. Energy data were not available for tests designated SEWA9806 and SEWA9807, thus they are not plotted in Figure 6.

A radar system was also employed during selected tests in this study, to measure the actual falling velocity of the safety hammers. Kinetic energy delivered at impact should theoretically be the half of the product of the falling mass and the square of the velocity at impact. For the limited data set produced in this field study, the energy at impact as measured by the BC Hydro transducer system was generally slightly higher than the energy calculated from velocity of falling safety hammers. Radar data were retained for future analysis, but were not rigorously processed as of the date of this paper.

Shear wave velocities measured between PVC-cased borings spaced about 10 ft apart at the Seward abutment site are plotted with depth in Figure 7. Shear wave velocities as measured were adjusted for the influence of effective overburden stress and replotted as hollow circles for comparison to published correlations, according to the following expression (NCEER, 1997):

$$V_{s1} = V_s(P_a / \sigma'_v)^{0.25} \tag{3}$$

where V_{s1} is stress-adjusted velocity, V_s is measured velocity, and P_a and σ'_v are defined as earlier. The groundwater table was assumed to be 10 ft (3.05 m) below ground surface, saturated and moist unit weights of the deposit were assumed to be 145 pcf (22.8 kN/m³) and 125 pcf (19.6 kN/m³), respectively, for the adjustment.

The stress-adjusted shear wave velocities given in Figure 7 are generally higher than those expected in liquefiable sandy soils; the NCEER (1997) report, for example, maintains that the upper bound stress-adjusted shear wave velocity for sandy gravels of Holocene age to be considered liquefiable is 220 m/s. The lowest adjusted velocity at the Seward site (about 200 m/s) was found in this study at 50 ft (15.2 m) depth. The soils sampled at this depth have the highest fines contents in the profile; the NCEER (1997) report recommends a stress-adjusted shear wave velocity bound less than 200 m/s for fine-grained, uncemented deposits.

CONCLUSIONS

A new, if limited, set of field data was produced from penetration tests conducted in gravels at a site that exhibited liquefaction behavior consequent to the 1964 Alaska Earthquake. These data represent the first collection of a variety of types of penetration tests and shear wave velocity measurements in gravels associated with liquefaction behavior where energy was measured to support adjustment of the results for varying equipment and procedures. Subsequent, more rigorous evaluation of the data will follow, in support of a research program to develop improved field procedures and interpretation methods for in situ determination of cyclic strength and behavior of coarse-grained soils. The data set is maintained by the principal author and may be made available to interested researchers in the near future.

If the gravels at the Seward site did, in fact, liquefy during the 1964 Alaska Earthquake, then either some factor exacerbated pore pressure buildup and strength loss in the gravelly soils or perhaps liquefaction was contained within the siltier gravels and sands found at depths of about 25 ft (7.6 m) and 50 ft (15.2 m). Although not presented herein, it is widely known that the 1964 ground motions were also quite severe and of long duration.

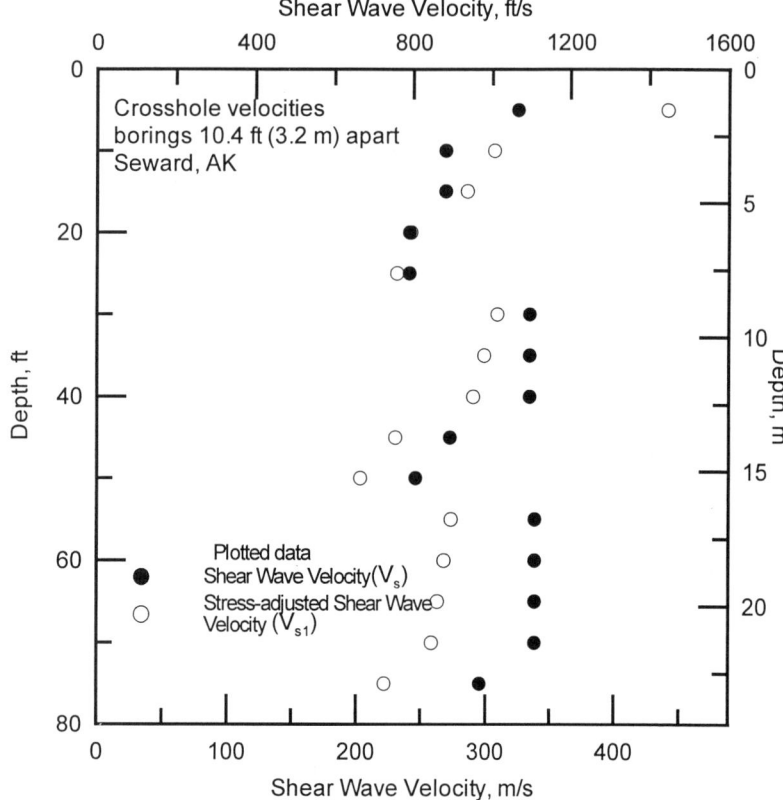

Figure 7. Crosshole Shear Wave Velocities Measured at Seward Test Site 1

Penetration resistance and shear wave velocity measurements at the Seward site do not indicate large zones of extraordinarily loose soils or other anomalous deposit characteristics, which presents somewhat of an enigma with regard to the widespread ground damage observed consequent to the 1964 event. Residents of the area recall stories of pervasive settlement of large expanses of ground near the test site; substantial densification may have resulted from earthquake shaking of the strength and duration attributed to the 1964 event. The authors believe this study will stimulate discussion in the in situ testing community, and hope that practitioners will critically examine the findings presented in this paper and communicate their reactions at the conference and among the profession.

ACKNOWLEDGMENTS

The tests described and the resulting data presented herein, unless otherwise noted, were obtained from research conducted under the Earthquake Engineering Research Program of the United States Army Corps of Engineers. Permission was granted by the Chief of Engineers to publish this information. The authors are grateful to the participants in the field study, namely: Messrs. Dan Ackerman and Erick St. Clair, the drilling equipment operators, from the US Army Engineer District, Alaska (CEPOA), whose conscientious and patient attention to a variety of demanding details assured the integrity of the data; Mr. Frank Dieckgraeff, owner of Metco, Inc., for site access and an abundant water supply; BC Hydro and Klohn-Crippen Consultants, Ltd. of Vancouver, BC, for the loan of the energy measurement system and radar devices, respectively; and the financial support and participation of Conetec Investigations, Ltd. and Dr. W. D. Liam Finn of Vancouver, BC, and Prof. John Howie of the Civil Engineering Department of the University of British Columbia, Vancouver, BC. The authors wish to particularly thank Professor Howie for his planning of the penetration test series and design of push rod subassemblies and adapters and cone penetrometers that were machined under the direction of CEPOA.

REFERENCES

American Society for Testing and Materials (ASTM) 1991. "Standard Test Methods for Crosshole Seismic Testing," D4428/D4428M-91, ASTM, West Conshohocken, PA.

Coulter, H.W. and Migliaccio, R.R. 1966. "Effect of the earthquake of March 27, 1964 at Valdez, Alaska, U.S. Geological Survey Professional Paper 542-C.

Harder, L.F. 1988. "Use of penetration tests to determine the liquefaction potential of soils during earthquake shaking," PhD Thesis, University of California, Berkeley, CA.

Harder, L.F. and Seed, H.B. 1986. "Determination of penetration resistance for coarse-grained soils using the Becker hammer drill," Report No. UCB/EERC-86/06, University of California, Berkeley, CA.

Ishihara, K. 1985. "Stability of natural slopes during earthquakes," *Proceedings, 11th International Conference on Soil Mechanics and Foundation Engineering,* San Francisco, CA, Vol. 1, 321-376.

Kokusho, T., Tanaka, Y., Kudo, K. And Kawai, T. 1995. "Liquefaction case study of volcanic gravel layer during 1993 Hokkaido-Nansei-Oki earthquake," *Proceedings, 3rd International Conference on Recent Advances in Geotechnical Earthquake Engineering and Soil Dynamics,* St. Louis, MO, 235-242.

Maurenbrecher, P.M., Den Outer, A. And Luger, H.J. 1995. "Review of geotechnical investigations resulting from the Roermond April 13, 1992 earthquake," *Proceedings, 3rd International Conference on Recent Advances in Geotechnical Earthquake Engineering and Soil Dynamics,* St. Louis, MO, 645-652.

McCulloch, D.S. and Bonilla, M.G. 1970. "Effects of the earthquake of March 27, 1964, on the Alaska Railroad," Geological Survey Professional Paper 545-D, U.S. Department of the Interior, Washington, D.C.

National Center for Earthquake Engineering Research (NCEER, now the Multidisciplinary Center for Earthquake Engineering Research, MCEER) 1997. "Proceedings of the NCEER Workshop on Evaluation of Liquefaction Resistance of Soils," edited by T. L. Youd and I. M. Idriss, Technical Report NCEER-97-0022, State University of New York at Buffalo, Buffalo, NY.

Sy, A. 1993. "Energy measurements and correlations of the Standard Penetration Test (SPT) and the Becker Penetration Test (BPT)," PhD Thesis, Department of Civil Engineering, University of British Columbia, Vancouver, BC.

Sy, A., and Campanella, R.G. 1993. "Dynamic performance of the Becker Hammer drill and penetration test," Canadian Geotechnical Journal, 30(4): 607-619.

Sy, A., and Campanella, R.G. 1994. "Becker and standard penetration tests (BPT-SPT) correlations with consideration of casing friction," Canadian Geotechnical Journal, 31(3): 343-356.

Tanaka, Y., Kokusho, T., Yoshida, Y., and Kudo, K. 1989. "Dynamic strength evaluation of gravelly soils," Proceedings of Discussion Session on Influence of Local Conditions on Seismic Response, Twelfth International Conference on soils Mechanics and Foundation Engineering, International Society of Soil Mechanics and Foundation Engineers, Rio de Janeiro, Brazil, 1989.

Tokimatsu, K. And Yoshimi, Y. 1983. "Empirical correlation of soil liquefaction based on SPT –value and fines content," Soils and Foundations, 23(4): 56-74.

Wang, W.S. 1984. "Earthquake damages to earth dams and levees in relation to soil liquefaction and weakness in soft clays," Proceedings of the International Conference on Case Histories in Geotechnical Engineering, St. Louis, MO, Vol. 1, 511-521.

Yegian, M.K., Ghahraman, V.G. and Harutiunyan, R.N. 1994. "Liquefaction and embankment failure case histories, 1988 Armenian earthquake," Journal of Geotechnical Engineering, ASCE, 120(3): 581-596.

Youd, T.L. , Harp, E.L., Keefer, D.K. and Wilson, R.C. 1985. "The Borah Peak, Idaho earthquake of October 28, 1983 - liquefaction," Earthquake Spectra, EERI, 2(1): 71-89.

Youd, T. L., and Idriss, I.M. eds. 1997. "Proceedings of the NCEER Workshop on Evaluation of Liquefaction Resistance of Soils," Report No. NCEER-97-0022, State University of New York at Buffalo, Buffalo, NY, viewable on the World Wide Web at http://mceer.buffalo.edu., 1997.

CALIBRATION OF SPT- AND CPT-BASED LIQUEFACTION EVALUATION METHODS

Caroline J. Chen[1] and C. Hsein Juang[2], M., ASCE

ABSTRACT

In this study, existing and newly developed SPT- and CPT-based simplified methods for liquefaction potential evaluation are calibrated with field performance databases. The calibration is carried out through probabilistic analyses of all cases in the associated database, which results in risk-based methods for liquefaction potential evaluation. The risk-based approach is applied to the SPT- and CPT-based databases from the 1989 Loma Prieta earthquake published recently by the U.S. Geological Survey.

INTRODUCTION

Because of the difficulty and the cost in obtaining high quality undisturbed soil samples, simplified methods based on in situ tests are preferred by geotechnical engineers for evaluating earthquake-induced liquefaction potential. These simplified methods are all developed from some database of field liquefaction performance records at sites that have been characterized with in situ tests such as standard penetration test (SPT), cone penetration test (CPT), and shear wave velocity measurements. The SPT-based simplified procedure, originally developed by Seed and Idriss (1971) and updated several times (e.g., Seed et al., 1985; Youd and Idriss, 1997), has become the standard of practice in North America and throughout much of the world. Most existing CPT-based simplified methods, such as Robertson and Wride (1998) and Stark and Olson (1995), are basically an extension of the Seed and Idriss method from standard penetration testing to cone penetration testing. Olsen (1997) presented a CPT chart-based method that is an update of Olsen (1984). Recently, Juang and Chen (1999a, b) proposed a new method for calculating liquefaction resistance based on their neural network modeling and analysis of field performance databases.

---

[1]Senior Staff Engineer, URS Greiner Woodward Clyde, 2020 East First St., Suite 400, Santa Ana, CA 92705.
[2]Professor, Department of Civil Engineering, Clemson University, Clemson, SC 29634-0911. E-mail: hsein@clemson.edu.

In the present study, two SPT-based methods, the SPT-SI method (the Seed and Idriss method as updated in Seed et al., 1985 and Youd and Idriss, 1997) and the SPT-JC method (Juang and Chen, 1999b), and two CPT-based methods, the CPT-Olsen method (Olsen, 1997) and the CPT-JC method (Juang and Chen, 1999a), are calibrated to field observations. These methods are all deterministic methods with which factor of safety against liquefaction (F_S) may be calculated and design decision may be made based on calculated F_S. In the present study, these methods are "calibrated" to field observations by means of mapping functions that mapped calculated F_S to the probability of liquefaction that is based field observations.

The developed mapping functions are examined using an additional field performance database published recently by the U.S. Geological Survey (Toprak et al., 1999). The results of the analyses are presented in this paper.

REVIEW OF DETERMINISTIC METHODS

A brief review of the four methods that are calibrated in the present study is in order. In a deterministic approach, the load and the resistance are calculated, followed by the calculation of factor of safety (F_S). In the liquefaction evaluation, the seismic load is generally expressed as cyclic stress ratio (Seed and Idriss, 1971; Seed et al., 1985):

$$CSR_{7.5} = 0.65 \left(\frac{\sigma_v}{\sigma_v'} \right) \left(\frac{a_{max}}{g} \right) (r_d) / MSF \qquad (1)$$

where σ_v is the total vertical stress at depth in question, σ_v' is the effective stress at the same depth, a_{max} is the peak horizontal ground surface acceleration, g is the acceleration due to gravity, MSF is the magnitude scaling factor, and r_d is the stress reduction factor. In their neural network modeling, Juang and Chen (1999a) rearrange Eq. (1):

$$CSR_{7.5} = 0.65 \ (R_p) \ (r_d) \ (SL) \qquad (2)$$

where R_p is a stress ratio ($R_p = \sigma_v/\sigma_v'$), SL is a combined seismic load parameter (SL = $a_{max}/g/MSF$). The term MSF is used to adjust the calculated CSR to the reference level of $M_w = 7.5$. Idriss (1999) proposed MSF be estimated using the following relationship:

$$MSF = 37.9 \ (M_w)^{-1.81} \quad \text{for } M_w \geq 5.75$$
$$= 1.625 \quad \text{for } M_w < 5.75 \qquad (3)$$

where M_w is the moment magnitude of the earthquake. The term r_d provides an approximate correction for flexibility of the soil profile. For a depth z of less than 23 m, the term r_d may be calculated using the following equation (Liao et al, 1988):

$$r_d = 1.0 - 0.00765z \quad \text{for } z \leq 9.15 \text{ m}$$
$$= 1.174 - 0.0267z \quad \text{for } 9.15 \text{ m} < z \leq 23 \text{ m} \qquad (4)$$

The liquefaction resistance of a soil is generally expressed as cyclic resistance ratio (CRR) and has been adopted in a recent National Center for Earthquake Engineering Research (NCEER) workshop (Youd and Idriss, 1997). In principle, CRR is equal to *critical* CSR, the maximum level of CSR a soil can resist without liquefaction. In the SPT-SI method, *critical* CSR or CRR is established by selecting a boundary curve that separates liquefied data points from non-liquefied data points in a plot of CSR versus corrected, clean-sand-equivalent SPT blow count, $(N_1)_{60cs}$. As in any deterministic simplified methods where the boundary curves are drawn to separate the zone of liquefaction from the zone of non-liquefaction, the *critical* CSR or CRR is *calibrated*, implicitly, by the field data. Thus, the question of whether Eqs. (3) and (4) are the most appropriate formulae to use in the calculation of CSR (since a number of other formulae are available) is *relatively* unimportant, provided that these methods are calibrated with adequate field performance data.

Whereas the simplified chart is still preferred by many geotechnical engineers as the way to represent the SPT-SI method, the following equation, established by Thomas Blake (Youd and Idriss, 1997), may be used for determining CRR:

$$CRR_{7.5} = \frac{a + cx + ex^2 + gx^3}{1 + bx + dx^2 + fx^3 + hx^4} \tag{5}$$

where a = 0.048, b = −0.1248, c = −0.004721, d = 0.009578, e = 0.0006136, f = −0.0003285, g = −0.00001673, and h = 0.000003741. The variable x in Eq. (5) is $(N_1)_{60cs}$, which is a function of the corrected SPT blow count, $(N_1)_{60}$, and the fines content (Youd and Idriss, 1997). This equation is valid for $(N_1)_{60,cs}$ less than 30. In the present study, the SPT-SI method is implemented by Eqs. (1), (3), (4), and (5).

Juang and Chen (1999b) developed a simplified equation for CRR based on their neural network analysis of field observations:

$$CRR_{7.5} = 0.241 \{\exp[(0.032+0.004FCI) (N_1)_{60}]\} -0.182 \tag{6}$$

where FCI is an index of fines content (FC) defined as follows: FCI = 1 for FC≤5%, FCI =2 for 5%<FC≤12%, FCI =3 for 12%<FC≤35%, and FCI = 4 for FC>35%. Use of ordinal scale to characterize the effect of fines content is consistent with current geotechnical knowledge. Equation (6) is established based on an extensive neural network modeling and analysis of a database of 233 field performance cases where SPT data are available (Juang and Chen, 1999b). The reader is referred to Juang and Chen (1999b) for a comparison of Eqs. (5) and (6). In the present study, the SPT-JC method is implemented by Eqs. (2), (3), (4), and (6).

The other two methods that are calibrated in the present study are CPT-based methods. The first one is the CPT-Olsen method. This method is a simplified version of the chart-based method by Olsen (1997). CRR is determined as follows:

$$CRR_{7.5} = 0.00128[q_c/(\sigma_v')^{0.7}] - 0.025 + 0.17R_f - 0.028R_f^2 + 0.0016R_f^3 \quad (7)$$

where q_c is the cone tip resistance in atm (1 atm is approximately equal to 100 kPa), σ_v' is the effective stress in atm, and R_f is the friction ratio in percent, defined as the sleeve friction f_s divided by q_c and multiplied by 100%. In the present study, the CPT-Olsen method is implemented by Eqs. (1), (3), (4), and (7).

The other CPT-based method considered in the present study, the CPT-JC method, is an artificial neural network (ANN) model developed by Juang and Chen (1999a). This ANN model was developed based on a database of 225 field performance cases compiled by Juang and Chen (1999a). CRR is calculated as follows:

$$CRR_{7.5} = f_T\left\{B_o + \sum_{k=1}^{n}\left[W_k \cdot f_T\left(B_{Hk} + \sum_{i=1}^{m} W_{ik}P_i\right)\right]\right\} \quad (8)$$

In this equation, the input variables are $P_1 = q_{cN1}$, $P_2 = R_f$, $P_3 = \sigma'_v$, and $P_4 = R_p$. The term q_{cN1} is the normalized cone tip resistance (Robertson and Wride, 1998):

$$q_{cN1} = q_c/\sigma'_v{}^{0.5} \quad (9)$$

where q_c is the measured cone tip resistance; both q_c and σ'_v must be in the unit of atm (which is 100 kPa). The coefficients B_0, B_{Hk}, W_k, and W_{ik} are obtained from neural network training (Juang and Chen, 1999a). The symbol f_T represents a transfer function defined below:

$$f_T(\theta) = 1 / (1 + e^{-\theta}) \quad (10)$$

Determination of CRR using Eq. (8) is straightforward, just as using the three methods (Eqs. 5, 6, and 7). However, the calculation could be tedious and a spreadsheet macro has been prepared (Juang and Chen, 1999a) and is available from the writers. In the present study, the CPT-JC method is implemented by Eqs. (2), (3), (4), and (8).

CALIBRATION OF THE SPT-JC METHOD

In a deterministic approach, CSR and CRR may be calculated using the models reviewed above, and the factor of safety F_S can then be determined. However, selection of a proper factor of safety to use in the design is not straightforward. Whereas experience provides a guide for the selection, a more rational approach should be based on calculated risk (or probability of liquefaction). Furthermore, use of risk-based approach enables a direct comparison of the results (in terms of probability) obtained from the four methods, whereas the factor of safety alone may not yield a meaningful comparison.

In the present study, the four deterministic methods are calibrated using a reliability-based procedure proposed by Juang et al. (1999). The calibration process yields a mapping function that maps calculated F_S to the probability of liquefaction. Using the SPT-JC method as an example, the process and the results of applying this procedure are described below.

Reliability Analyses

The reliability index, β, is calculated using second moment methods. Specifically, the Hasofer-Lind reliability index is computed (Ditlevsen, 1981):

$$\beta = \min_{x \in F} \sqrt{(X - m)^T C^{-1} (X - m)} \tag{11}$$

where X = the vector of random variables in the constraint function given by $G(X) = 0$,
m = the vector of mean values,
C = the covariance matrix.

The minimization in Eq. (11) is performed over the failure domain F corresponding to the region $G(X) < 0$. In a reliability analysis of liquefaction potential, the constraint may be written $G(X) = CRR/CSR - 1 = 0$, where CRR may be determined by the SPT-JC model (Eq. 6) and CSR may be determined by Eq. (2). In the present study, Monte Carlo simulation technique is used to find the minimum β that satisfies the constraint.

In the CSR model (Eq. 2), the variable r_d is a function of depth and it is not considered as a random variable in the reliability analysis. The variable R_p is a function of σ_v and σ_v' and the variable SL is a function of a_{max} and M_w. The four basic variables (σ_v, σ_v', a_{max} and M_w) are treated as random variables in the reliability analysis. According to Juang et al. (1999), the coefficients of variation (COV) for σ_v and for σ_v' may be taken at 0.10 and 0.15, respectively. The COV for M_w, the measured earthquake magnitude, is taken as 0.05 (Espinosa, 1982; Comartin et al, 1995), while the COV for a_{max} is taken as 0.20, which is about the average of the reported values at individual sites in the database (Juang and Chen, 1999b). These input variables are assumed to follow normal distribution. A sensitivity analysis of the effect of these assumptions was conducted by Juang and Chen (1999a,b), and no significant effect on the obtained mapping functions was found.

In the CRR model (Eq. 6), two input variables, $(N_1)_{60}$ and FCI are used. The parameter FCI is an index and is not treated as a random variable herein. The variable $(N_1)_{60}$ is derived from σ_v' and N_{60}, the SPT N value without overburden pressure correction. Thus, only one additional variable N_{60} is treated as random variable in the reliability analysis. Based mostly on Kulhawy and Trautmann (1996), the COV for N_{60} is assumed to be 0.30. The variable N_{60} is assumed to follow normal distribution.

Note that in the present reliability analysis, no uncertainty is assigned to the CSR model (Eq. 2) *itself*, since it is used as a "reference" in the development of the CRR model (Juang and Chen, 1999b). In other words, the effect of the uncertainty associated with the CSR model itself is built into the CRR model. Furthermore, the uncertainty of the CRR model is excluded from the present reliability analysis but for different reasons. First, it is difficult to quantify this model uncertainty. Second, whatever error exists in the model will be accounted for in the calibration, as the calculated reliability indexes are subsequently *calibrated* to the probability of liquefaction based on field performance records in the database.

Another issue warranting consideration in the reliability analysis is the possible correlation among the input variables. There is strong correlation between σ_v and σ_v', and between a_{max} and M, while the correlations for all other pairs are rather weak (Juang et al., 1999). Based on available data, the average correlation coefficient between σ_v and σ_v' is determined to be about 0.95, and the average correlation coefficient between a_{max} and M is determined to be about 0.90. These values are used in the present study, whereas all other pairs of variables are assumed to be independent.

The database of 233 cases (Juang and Chen, 1999b), consisting of 145 liquefied cases and 88 non-liquefied cases, is used in the reliability analyses. Details of the calculation of reliability index β using Eq. (11) are documented in Juang and Chen (1999b).

Distribution of Reliability Index

For each case in the database, a reliability index β is obtained. The calculated β values are then grouped according to whether or not liquefaction actually occurred at the site. Figure 1 shows the histograms of the calculated reliability indices for all cases. An evaluation of the distribution of β values reveals that both Group L (liquefied cases) and Group NL (non-liquefied cases) can be fitted well with normal distribution:

$$f(x) = \frac{1}{\sqrt{2\pi\sigma^2}} \exp\left[-\frac{(x-\mu)^2}{2\sigma^2}\right] \tag{12}$$

where μ is the mean and σ is the standard deviation. For Group NL, $\mu = 0.77$ and $\sigma = 1.05$, and for Group L, $\mu = -0.82$ and $\sigma = 1.16$. The derived distributions are considered appropriate, although further investigation into this issue is warranted, when more field case records become available.

Mapping Reliability Index to Probability of Liquefaction

Logistic regression methods have been used to analyze the probability of liquefaction in conjunction with the Seed and Idriss method (Liao et al., 1988; Toprak et al., 1997). In this study, a different approach is taken. Here, conditional probability that liquefaction

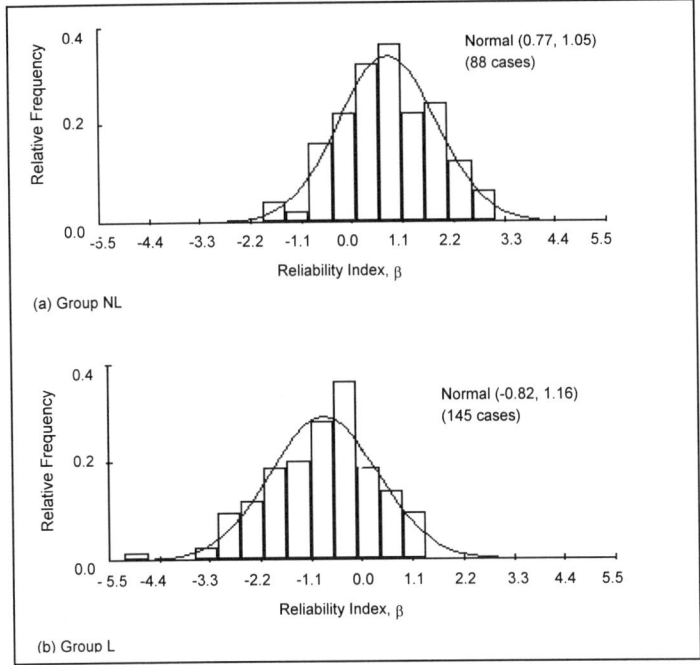

Figure 1. Histograms of Reliability Indices calculated with the SPT-JC Method

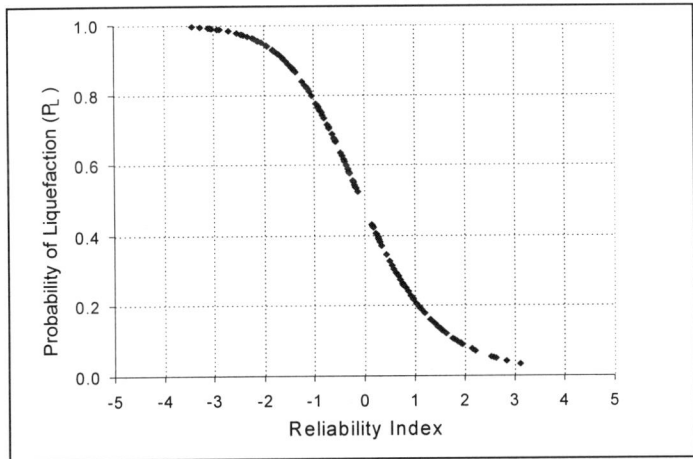

Figure 2. Mapping Reliability Index to Probability of Liquefaction (SPT-JC Method)

will occur for a case in which a reliability index β has been calculated can be determined according to Bayes' theorem:

$$P(L|\beta) = \frac{P(\beta|L)\,P(L)}{P(\beta|L)\,P(L) + P(\beta|NL)\,P(NL)} \tag{13}$$

where $P(L|\beta)$ = the probability of liquefaction for a given β,
$P(\beta|L)$ = the distribution function of β, given that liquefaction did occur,
$P(\beta|NL)$ = the distribution function of β, given that liquefaction did not occur,
$P(L)$ = prior probability of liquefaction,
$P(NL)$ = prior probability of no-liquefaction.

If knowledge of prior probabilities $P(L)$ and $P(NL)$ is available, Eq. (13) can be used to determine the probability of liquefaction for a given β. In the absence of knowledge about the prior probabilities, it may be assumed, based on the principle of maximum entropy, that $P(L) = P(NL)$. Under the assumption that $P(L) = P(NL)$, Eq. (13) can be rewritten (Juang et al., 1999):

$$P(L|\beta) = \frac{f_L(\beta)}{f_L(\beta) + f_{NL}(\beta)} \tag{14}$$

where $f_L(\beta)$ and $f_{NL}(\beta)$ are the probability density functions of β for Groups L and NL, respectively (see Figure 1). In the probabilistic analyses presented herein, Eq. (14) is used to determine the probability of liquefaction for all cases with calculated β values. The result, expressed as a mapping from reliability index to probability of liquefaction, is shown in Figure 2.

Mapping Factor of Safety to Probability of Liquefaction

Figure 3 shows the relation between F_S and P_L based on the 233 cases studied. Each data point corresponds to a particular field case, where P_L is the probability of liquefaction calculated using Eq. (14), in which β is calculated by Eq. (11), and $F_S = CRR/CSR$. Considering all points equally (whether they belong to Group NL or Group L), a best-fit curve represented by the following equation is obtained ($R^2 = 0.99$):

$$P_L = 1 / [1 + (F_S/a)^b] \tag{15}$$

where a = 1.0, and b = 3.50. This mapping function calibrates a deterministic method (the SPT-JC method in this case) to the field performance records so that for a calculated F_S value, the probability of liquefaction is inferred. This mapping function provides a basis for a risk-based decision using a deterministic approach. In other words, a design decision in terms of selection of a proper factor of safety can be made based on an acceptable risk (i.e., probability of liquefaction).

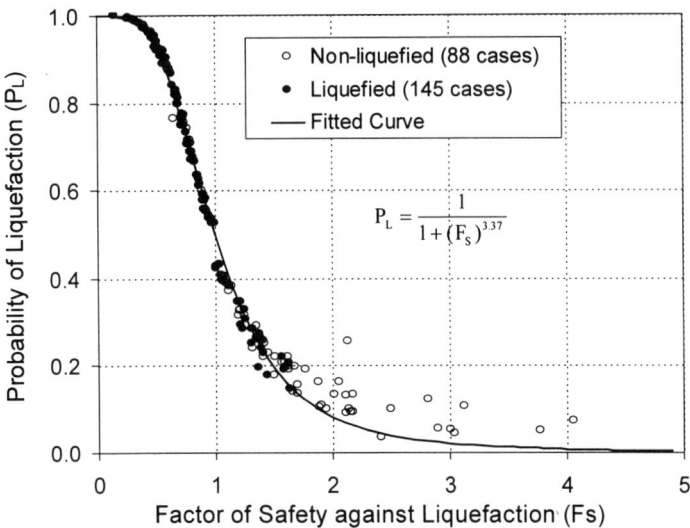

Figure 3. P_L-FS Mapping Function Derived Based on Equation 14 (SPT-JC Method)

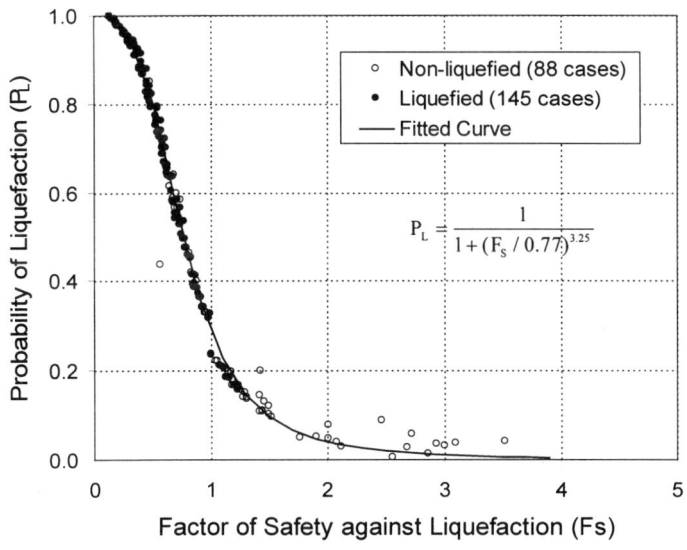

Figure 4. P_L-FS Mapping Function Derived Based on Equation 14 (SPT-SI Method)

CALIBRATION OF SPT-SI, CPT-Olsen, CPT-JC METHODS

Following the same procedure as described above, the SPT-SI method is calibrated using the same database of 233 field performance cases. Figure 4 shows the mapping functions that mapped the calculated factor of safety to the probability of liquefaction. This mapping function, which calibrates the SPT-SI method, has the same general form as Eq. (15). For the SPT-SI method, a = 0.77, and b = 3.25. Note that from Fig. 4, the probability of liquefaction, P_L, is about 0.30 at a calculated F_S of 1.0. This probability agrees well with the results obtained from logistic regression methods (e.g., Liao et al., 1988; Toprak et al., 1999).

Calibration of the CPT-based methods can be carried out following the same approach (Eq. 14). The resulting mapping functions that calibrate the four deterministic methods are shown in Figure 5, while the coefficients in these functions are listed in Table 1. In the SPT-SI method, the boundary curve that separates liquefaction from non-liquefaction, represented by F_S = 1, is characterized by a probability of 30%, whereas the other three methods are shown to have a probability of 50% at F_S = 1.

Table 1. Summary of Mapping Functions Obtained in This Study

Method	Mapping Function (see Eq.15)
SPT-SI	a = 0.77, b = 3.25
SPT-JC	a = 1.0, b = 3.37
CPT-Olsen	a = 1.0, b = 2.78
CPT-JC	a = 1.0, b = 4.65

It is noted that the calibration of a deterministic method might also be carried out by calculating the factor of safety for each case, and then plotting the histogram of the calculated F_S values. With this approach, the probability of liquefaction may be determined for a given F_S:

$$P(L|F_S) = \frac{f_L(F_S)}{f_L(F_S) + f_{NL}(F_S)} \tag{16}$$

Applying this approach to calibrate the SPT-SI method yields practically the same mapping function (a= 0.77 and b= 3.21) as that obtained through the reliability analyses and shown in Table 1. Whereas this result is interesting, further investigation into this result and the F_S-based approach taken is warranted.

CASE STUDY

Following the 1989 Loma Prieta, California earthquake, the USGS compiled extensive SPT and CPT at 25 sites in the Monterey and San Francisco Bay regions. Sites with and without evidences of liquefaction were investigated. Table 2 lists 48 sites where field observations along with both SPT and CPT measurements were made. These data

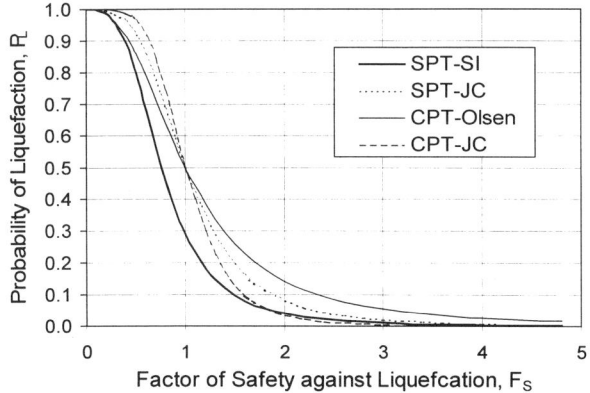

Figure 5. Mapping Functions Established for the Four Methods

are taken directly from Toprak et al. (1999) and are identical to those listed in Toprak et al. (1999) except that two sites where CPT data are not available are excluded from consideration. In the present study, these sites are analyzed using the four methods presented above and the probability of liquefaction at each site is calculated using the associated mapping functions that calibrate these methods.

The calculated probability of liquefaction for each of the 48 cases using each of the four methods is also listed in Table 2. As a way to interpret the calculated probability, the likelihood of the occurrence of liquefaction is divided into 5 classes:

Probability	Class	Description (likelihood of liquefaction)
$0.85 \leq P_L$	5	Almost certain that it will liquefy
$0.65 \leq P_L < 0.85$	4	Very likely
$0.35 \leq P_L < 0.65$	3	Liquefaction/non-liquefaction is equally likely
$0.15 \leq P_L < 0.35$	2	Unlikely
$P_L < 0.15$	1	Almost certain that it will not liquefy

To compare the four methods, success rate is calculated for each method (Table 3). Three criteria are used to define "success" in the prediction for liquefied cases: Criterion A, $P_L \geq 0.85$; Criterion B, $P_L \geq 0.65$; and Criterion C, $P_L \geq 0.15$. Criterion A is the most stringent, considering as a success only if it belongs to Class 5; Criterion B considers both Classes 4 and 5 as a success; and Criterion C considers any classes other than Class 1 as a success. Similarly, three criteria are used to define success for non-liquefied cases: Criterion D, $P_L < 0.15$; Criterion E, $P_L < 0.35$; and Criterion F, $P_L < 0.85$. Criterion D is the most stringent, considering as a success only if it belongs to Class 1; Criterion E considers both Classes 1 and 2 as a success; and Criterion F considers any classes other than Class 5 as a success.

Table 2. SPT and CPT Database from the 1989 Loma Prieta Earthquake (Mw = 6.9), Source: Toprak et al. (1999)

Borehole Location	Case No.	L	D (m)	σ_v (kPa)	σ_v' (kPa)	FC (%)	$(N_1)_{60}$	q_c (kPa)	R_f	a_{max} (g)	Calculated Probability of Liquefaction			
											SPT-SI	SPT-JC	CPT-Olsen	CPT-JC
AIR-18	1	1	4.3	82.4	63.7	21	7	6196	0.39	0.26	0.71	0.70	0.74	0.81
AIR-21	2	1	4.6	88.2	66.7	7	4.1	6723	0.43	0.26	0.96	0.89	0.69	0.80
CMF-3	3	1	5.8	109.1	108.1	41	6.4	2398	0.54	0.36	0.69	0.73	0.93	0.99
CMF-8	4	1	6.0	113.7	102.9	17	11.1	8362	0.65	0.36	0.65	0.52	0.72	0.76
FAR-58	5	1	7.6	145.1	117.7	5	19.5	8658	0.46	0.36	0.51	0.30	0.86	0.88
FAR-59	6	1	5.2	98.1	94.2	16	10.3	2794	0.86	0.36	0.67	0.53	0.82	1.00
FAR-61	7	1	6.5	124	101.5	21	11.1	4222	0.83	0.36	0.68	0.60	0.85	0.99
GRA-123	8	1	7.5	143	118.5	18	9.2	4586	0.41	0.34	0.73	0.66	0.94	0.96
JEF-121	9	1	5.8	111	87.4	7	3.6	6816	0.44	0.21	0.92	0.80	0.59	0.69
JEF-141	10	1	3.7	70.8	55.1	26	6.6	1802	0.83	0.21	0.52	0.57	0.66	0.90
JEF-148	11	1	7.3	140.7	98.5	5	10	7687	0.38	0.21	0.68	0.48	0.69	0.35
JEF-149	12	1	5.8	111.3	83.8	2	6.1	4839	0.50	0.21	0.87	0.70	0.71	0.74
JEF-32	13	1	2.4	45.6	39.7	4	14.1	2226	0.40	0.21	0.29	0.21	0.79	0.77
KET-74	14	1	2.7	51.7	40.0	15	21.8	4923	1.20	0.47	0.54	0.38	0.75	0.99
LEN-39	15	1	4.3	82.8	59.2	11	10.1	3670	0.19	0.22	0.62	0.47	0.93	0.61
LEN-51	16	1	3.4	65.2	49.6	36	2.8	1156	0.43	0.22	0.79	0.87	0.94	0.63
LEN-53	17	1	3.4	65	52.2	10	8.2	3431	0.20	0.22	0.72	0.52	0.90	0.77
ML-15	18	1	1.5	28.2	28.2	46	9.3	1040	1.35	0.28	0.32	0.45	0.47	0.98
MRR-65	19	1	7.6	144.5	124.9	12	14.5	6898	0.77	0.40	0.65	0.50	0.81	0.96
PD-44	20	1	5.8	111	87.4	3	13.1	7992	0.58	0.22	0.44	0.27	0.47	0.46
RAD-99	21	1	6.4	122.2	99.6	18	10	4183	1.24	0.38	0.78	0.71	0.74	0.99
SEA-31	22	1	3.4	66	40.5	29	7.7	1160	0.52	0.22	0.62	0.74	0.94	0.66
SIL-68	23	1	6.2	118.7	92.2	15	9.1	4546	0.68	0.38	0.87	0.79	0.89	0.98
SIL-71	24	1	6.1	115.6	103.9	9	13.5	5486	0.62	0.38	0.70	0.49	0.85	0.99
SPR-45	25	1	4.8	90.6	86.6	6	13.2	4862	0.60	0.33	0.63	0.35	0.79	0.95
SPR-48	26	1	6.4	121.2	110.4	13	8.3	3696	0.81	0.33	0.78	0.64	0.78	0.99
TAN-103	27	1	5.0	94.3	90.4	22	8.2	4768	0.86	0.13	0.07	0.06	0.11	0.28
CCK-1	28	0	9.8	187.7	145.5	7	12.4	8630	1.52	0.17	0.27	0.11	0.13	0.01

Table 2. SPT and CPT Database from the 1989 Loma Prieta Earthquake (Continued)

Borehole Location	Case No.	L	D (m)	σv (kPa)	σv' (kPa)	FC (%)	(N1)60	qc (kPa)	Rf (%)	amax (g)	Calculated Probability of Liquefaction			
											SPT-SI	SPT-JC	CPT-Olsen	CPT-JC
CCK-16	29	0	7.6	146.8	98.7	28	21.3	4601	0.43	0.17	0.03	0.02	0.77	0.45
CCK-23	30	0	8.2	158.8	101.9	7	9.9	6865	1.72	0.17	0.59	0.32	0.18	0.01
CMF-10	31	0	7.6	146.6	101.4	20	14.8	7721	1.13	0.36	0.62	0.49	0.71	0.60
LEN-37	32	0	3.4	64.6	55.8	13	11.9	3004	0.37	0.22	0.33	0.19	0.83	0.81
LEN-52a	33	0	3.4	64.5	57.6	12	13	5595	0.54	0.22	0.25	0.17	0.44	0.48
M-2	34	0	7.0	132.1	90.4	4	11.8	5900	0.58	0.16	0.37	0.21	0.48	0.20
M-6	35	0	5.8	102.7	99.7	6	31.3	3860	0.54	0.16	0.00	0.00	0.44	0.65
MAR-110	36	0	3.7	69.1	51.5	13	5.5	3109	0.64	0.13	0.50	0.31	0.35	0.40
MAR-111	37	0	4.9	94.7	63.3	15	5.6	2662	0.45	0.13	0.50	0.37	0.67	0.22
MCG-136	38	0	2.7	51	48.1	11	8.5	3178	0.88	0.26	0.65	0.52	0.49	0.88
MCG-138	39	0	3.7	71.1	52.4	29	4.1	1692	0.77	0.26	0.84	0.89	0.83	0.93
ML-14	40	0	3.7	71.6	49.9	3	30.6	14778	0.10	0.28	0.00	0.06	0.38	0.01
MRR-67	41	0	6.4	120.5	118.5	15	28.6	14326	0.68	0.40	0.00	0.02	0.52	0.07
PD-43	42	0	3.7	70.4	59.7	19	24.2	7864	0.45	0.22	0.02	0.02	0.38	0.27
PD-79	43	0	4.9	94.8	61.5	3	31.3	4966	1.25	0.22	0.00	0.03	0.42	0.71
RAD-98	44	0	5.5	105	85.4	7	16.9	8998	0.51	0.38	0.65	0.39	0.78	0.75
SEA-29	45	0	7.8	151.3	94.4	8	41	17470	0.51	0.22	0.00	0.00	0.29	0.01
SRB-116	46	0	6.4	120.3	120.3	7	8	4510	2.88	0.11	0.16	0.04	0.01	0.00
SRB-117	47	0	6.4	120.3	120.3	13	8	4510	2.17	0.11	0.09	0.03	0.02	0.01
TAN-105	48	0	4.9	92.7	85.8	41	7.2	4007	0.47	0.13	0.08	0.08	0.38	0.32

Note: In column 3, L = 1 indicating liquefaction; L = 0 indicating no liquefaction. Abbreviations for sites listed below.

AIR = Airport; CMF = Clint Miller Farm; CCK = Coyote Creek; FAR = Farris Farm; GRA = Granite Construct. Co.
JEF = Jefferson Farm; KET = Kett; LEN = Leonardini; M = Marina District; MRR = Marinovich; MAR = Martella
MCG = McGowan; ML = Moss Landing; PD = Pajaro Dunes; RAD = Radovich; SRB = Salinas; SEA = Sea Mist
SIL = Silliman; SPR = South Pacific Bridge; TAN = Tanimura.

A quick examination of the 48 cases reveals that all four methods fail to classify two sites. At site TAN-103 where the field evidence suggests the occurrence of liquefaction, all four methods *strongly* indicated no liquefaction. At site MCG-138 where the field evidence suggests no liquefaction, all four methods *strongly* indicated the occurrence of liquefaction. Since the four methods were developed based either on different in situ tests (SPT versus CPT) or on different approach (traditional versus neural network model), the results seem to suggest that these two cases might be considered as outliers. Thus, they are *excluded* from further consideration in this paper.

Table 3 shows a comparison of the success rate according to each of the defined criteria. In all four methods, the success rate increases as the criterion for success becomes less stringent (i.e, for liquefied cases, from Criterion A to B and then to C; for non-liquefied cases, from Criterion D to E and then to F). On liquefied cases, all four methods yield a success rate of 100% if Criterion C (PL $\geq$ 0.15) is used. On non-liquefied cases, all four methods yield a success rate of 100% if Criterion F (PL < 0.85) is used. It is observed from Table 3 that the CPT-based methods have higher success rates in predicting liquefied cases, and lower success rates in predicting non-liquefied cases, than the SPT-based methods. Overall, the four methods are considered comparable in their accuracy of predicting the occurrence of liquefaction/non-liquefaction, although the CPT-JC method appears to be slightly more accurate than the other three methods based on *limited cases* studied. This result is understandable because all four methods have been calibrated using the same procedure and comparable database of field performance records.

Table 3. Success Rates in the Liquefaction Predictions Using the Four Methods

	SPT-SI	SPT-JC	CPT-Olsen	CPT-JC
Liquefied Cases				
Criterion A ($P_L \geq 0.85$)	16%	8%	34%	54%
Criterion B ($P_L \geq 0.65$)	69%	38%	88%	84%
Criterion C ($P_L \geq 0.15$)	100%	100%	100%	100%
Non-liquefied Cases				
Criterion D ($P_L < 0.15$)	45%	55%	15%	45%
Criterion E ($P_L < 0.35$)	65%	80%	30%	50%
Criterion F ($P_L < 0.85$)	100%	100%	100%	100%

As a final note, it should be kept in mind that the liquefaction classes defined above are for "prediction" of the occurrence of liquefaction/non-liquefaction. This is different from the design practice where a more conservative approach may be taken.

IMPLICATION OF THE CALIBRATION STUDY

Once a mapping function is established for a particular method, the required F_S can be obtained for a specified risk (probability of liquefaction). Thus, a design at this FS level will assure a probability of liquefaction of less than the specified risk. It follows that the same level of risk can be attained regardless of which method is used, provided that the method has been calibrated with a sufficiently large database and that a proper F_S is specified. For example, if a risk level of 20% is specified, the required F_S for the SPT-SI, SPT-JC, CPT-Olsen, and CPT-JC methods are 1.18, 1.48, 1.70, and 1.34, respectively.

A couple of conclusions may be inferred from the results presented above. First, among the four methods examined, the SPT-SI method is the most conservative, and thus it requires the least F_S to attain the same level of risk, whereas the CPT-Olsen method is the least conservative. Second, without the calibration of the methods to the probability of liquefaction based on field observations, comparison of different methods in the traditional deterministic approach is meaningless. For example, if the engineer adopts an F_S of, say, 1.3, the level of risk involved will be different using different methods.

CONCLUDING REMARKS

Existing and newly developed SPT- and CPT-based simplified methods for liquefaction potential evaluation have been calibrated with field performance databases. Mapping functions resulting from calibration of these deterministic methods have been established. These mapping functions, which relate calculated factor of safety to the probability of liquefaction based on field observations, provide a basis for risk-based design against the occurrence of liquefaction. All four methods may be used to evaluate liquefaction for a specified risk level, although the required factor of safety would be different among these methods. The case study using high quality data published by the U.S. Geological Survey has demonstrated the use and versatility of the calibrated methods. More case studies are needed to assess and compare the accuracy of the four methods examined.

ACKNOWLEDGMENTS

This study is supported by the National Science Foundation through Grant No. CMS-9612116. The program official for this NSF grant is Dr. Clifford Astill. This financial support is appreciated. Dr. Selcuk Toprak of the USGS is thanked for his clarification of the data set used in the case study. Dr. Ronald Andrus of Clemson University is thanked for his review of the manuscript.

REFERENCES

Comartin, C.D., Greene, M., and Tubbesing, S.K. (1995), *The Hyogoken-Nambu Earthquake Preliminary Reconnaissance Report*, EERI Report No. 95-40, Earthquake Engineering Research Institute, California.
Ditlevson, O. (1981), *Uncertainty Modeling*, McGraw Hill, New York.
Espinosa, A.F. (1982), M_L and M_o determination from strong-motion accelerograms, and expected intensity distribution. *Geological Survey Professional Paper 1254, The Imperial Valley, California, Earthquake of October 15, 1979*, United States Government Printing Office, Washington, 1982.

Idriss, I.M. (1999), An update of the Seed-Idriss simplified procedure for evaluating liquefaction potential, *Proceedings, TRB Workshop on New Approaches to Liquefaction Analysis*, FHWA-RD-99-165, Federal Highway Administration, Washington, D.C.

Juang, C.H. and Chen, C. J. (1999a), *A CPT-based Method for Assessing Liquefaction Potential of Sandy Soils*, Technical Report, Department of Civil Engineering, Clemson University, Clemson, S.C., March 1999, 64p.

Juang, C.H. and Chen, C. J. (1999b), *Risk-Based Liquefaction Potential Evaluation Using SPT*, Technical Report, Department of Civil Engineering, Clemson University, Clemson, S.C., August 1999, 63p.

Juang, C. H., Rosowsky, D.V. and Tang, W.H. (1999), A reliability-based method for assessing liquefaction potential of sandy soils, *Journal of Geotechnical and Geoenvironmental Engineering*, ASCE, Vol. 125, No. 8, pp. 684-689.

Kulhawy, F.H. and Trautmann, C.H. (1996), Estimation of in situ test uncertainty, *Uncertainty in the Geologic Environment: From Theory to Practice*, GSP No. 58, ASCE, New York, pp. 269-286.

Liao, S.S.C., Veneziano, D., and Whitman, R.V. (1988), Regression model for evaluating liquefaction probability, *Journal of Geotechnical Engineering*, ASCE, Vol. 114, No.4, pp. 389-410.

Olsen, R.S. (1984), Liquefaction analysis using the cone penetration test, *Proceedings of the Eighth World Conference on Earthquake Engineering*, Vol. III, pp. 247-254, Prentice-Hall, Inc., Englewood Cliffs, NJ.

Olsen, R.S. (1997), Cyclic liquefaction based on the cone penetrometer test, *Proceedings of the NCEER Workshop on Evaluation of Liquefaction Resistance of Soils*, Technical Report NCEER-97-0022, T. L. Youd and I. M. Idriss, eds., State University of New York at Buffalo, Buffalo, NY, pp. 225-276.

Robertson, P.K. and Wride, C.E. (1998), Evaluating cyclic liquefaction potential using the cone penetration test, *Canadian Geotechnical Journal*, Vol. 35, No. 3, pp. 442-459.

Seed, H.B. and Idriss, I.M. (1971) Simplified procedure for evaluating soil liquefaction potential, *Journal of the Soil Mechanics and Foundation Div.*, ASCE, Vol. 97, SM9, pp. 1249-1273.

Seed, H.B., Tokimatsu, K., Harder, L.F., and Chung, R. (1985), "Influence of SPT procedures in soil liquefaction resistance evaluations," *Journal of Geotechnical Engineering*, ASCE, Vol. 111, GT 12, pp. 1425-1445.

Stark, T.D. and Olson, S.M (1995), Liquefaction resistance using CPT and field case histories, *Journal of Geotechnical Engineering*, ASCE, Vol. 121, No. GT 12, pp. 856-869.

Toprak, S., Holzer, T.L., Bennett, M.J., and Tinsley, J.C., III (1999), CPT- and SPT-based probabilistic assessment of liquefaction, *Proceedings of Seventh US-Japan Workshop on Earthquake Resistant Design of Lifeline Facilities and Counter-measures Against Liquefaction*, Seattle, August 1999, Multidisciplinary Center for Earthquake Engineering Research, Buffalo, NY.

Youd, T. L. and Idriss, I. M., eds. (1997), *Proceedings of the NCEER Workshop on Evaluation of Liquefaction Resistance of Soils*, Technical report NCEER-97-0022, T. L. Youd and I. M. Idriss, eds., State University of New York at Buffalo, NY.

SOIL STRATIGRAPHY DELINEATION BY VisCPT

Ali M. Ghalib[1], Roman D. Hryciw[2], Endra Susila[3]

ABSTRACT

The traditional CPT and CPTU tests are very good tools for stratigraphic soil logging. Using empirical techniques based on tip resistance, local friction and excess pore pressure, a continuous soil profile can be established which is more detailed than that obtained by a conventional SPT test and sampling. However, because tip resistances are controlled by the mechanical soil properties averaged over at least 5-10 cone diameters of elevation, and because the local friction is measured over a 13 cm distance, even the CPT cannot easily discern relatively thin lenses and seams in the stratigraphy. Such lenses may nevertheless control the hydrologic and mechanical behavior of a site. The vision cone penetrometer (VisCPT) collects a continuous stream of magnified images of the soil stratigraphy and thus, every lens and seam is identified. The present paper demonstrates that textural image indices show very high correlation to soil stratigraphy, as confirmed by high quality continuous borehole samples, while the CPT alone often misses or misinterprets the constituents of thin lenses and seams.

INTRODUCTION

The Cone Penetration Test (CPT) has become recognized as a valuable in-situ testing technique because of its speed, reliability, cost effectiveness and soil profiling capabilities. Important applications of the CPT in geotechnical engineering include site characterization and soil profiling, design of deep foundations, design and evaluation of ground improvement, assessment of liquefaction potential and estimation of soil stiffness and strength.

1. Assistant Project Engineer, NTH Consultants Ltd., Detroit, MI, 48226
2. Professor, Civil and Environmental Engineering Department, University of Michigan, Ann Arbor, MI 48109
3. Graduate Student, Civil and Environmental Engineering Department, University of Michigan, Ann Arbor, MI 48109.

In recent years, the CPT has also undergone significant development and modification to address geo-environmental issues. There have been important improvements in sensors, data acquisition and processing capability, and in the analytical interpretation of CPT results (Robertson et al., 1998). The additional sensors include electrical resistivity/conductivity, pH and ion detectors, gamma/neutron as well as special fiber optic devices for liquid and/or vapor detection. In geo-environmental applications, the CPT is used for the characterization of groundwater flow conditions, estimation of hydraulic conductivity, assessment of contaminant types and distributions and design of site remediation methods.

The CPT by comparison to other available penetration tests provides the most continuous, detailed profile of the soil strata. Using empirical techniques based on tip resistance, local friction and excess pore pressure, a continuous soil profile can be established which is more detailed than one obtained by a conventional standard penetration test (SPT) and sampling. Unfortunately, the CPT has one important deficiency namely, the absence of a recovered sample for direct examination of the soil. As such, a number of soil identification and classification charts have been proposed based on actual or normalized values of tip resistance, friction ratio and pore water pressure (Douglas and Olsen, 1981, Robertson et al., 1986, Robertson 1990, Olsen and Mitchell, 1995). These classification charts have demonstrated their effectiveness over the years. However, reliance on them in the absence of site specific validation of soil types by sampling or by prior experience with similar soil conditions is not recommended (Mitchell and Brandon, 1998). There have been many cases wherein reliance on the classifications given by the charts in the absence of ground truth would have yielded incorrect results (Suzuki et al, 1995, Arango, 1997).

The CPT friction ratio is known to generally increase with the increase of soil fines content and plasticity. However, for friction ratios between 0.25 and 1.0 percent, values typical of sandy soils, Suzuki et al. (1995) observed fines content ranging from 2% to 50%. From CPT tests performed in Indonesia, Dubai, San Francisco, New Mexico and Trinidad, Arango (1997) found that the fines content ranged from 8.5% to more than 90% for the same range of friction ratios. The lack of a strong fines content-friction ratio correlation can be attributed to friction ratio being dependent on factors other than fines content. These factors include the in-situ lateral stresses, nature of fines (plasticity), soil sensitivity, cone sleeve surface properties and the possible level and type of soil contamination.

The failure zone that develops around an advancing cone penetrometer extends radially and beneath the cone tip to a distance of 5-10 and 10-20 cone diameters for soft and stiff soils, respectively (Lunne et al. 1997). As such, the penetration resistance that is measured will be influenced by the strength and deformation properties of the soil at some distance from the cone. In fact, in the case of a dense sand layer underlying a soft clay layer, the sand thickness has to be 0.75 m or more for a standard CPT cone to record the full resistance expected of the sand. Thus, the measured value of tip resistance q_c may not give a true measure of the actual state of the soil at the elevation of the cone tip (Treadwell, 1976). The layer effect could also lead to incorrect soil identification if tip and friction ratio based classification charts are used. Based on elasticity theory, Vregudenhil et al. (1994) suggested a "thin layer correction factor, K_c" to estimate the tip resistance that would have developed in a sand layer sandwiched between two thick layers of soft soil, had the sand layer been thicker. However, the correction requires prior knowledge of the thickness of the sand layer, which itself is unknown without direct soil observations.

This paper presents an improved new technique for in-situ soil classification and thin layer delineation utilizing a Vision Cone Penetrometer (VisCPT) and image textural analysis. The next section briefly describes the VisCPT and the supporting digitization and data processing systems. The following section presents the method of image textural analysis as a means for soil classification. A quantitative correlation between the average grain size and three image textural indices will be presented with reference to a laboratory prepared layered soil sample. Finally the delineation capability of the pro-posed method will be evaluated for a thinly layered soil profile. A performance comparison with a standard CPT-based classification is presented.

THE VISION CONE PENETROMETER

The VisCPT was developed by Raschke and Hryciw (1997) for capturing continuous video images of a site's soil stratigraphy. By adapting a miniature CCD video cameras to a standard electronic cone penetrometer, the VisCPT has overcome the major shortcoming of the CPT vis-à-vis the Standard Penetration Test (SPT), that is, the inability to visually observe and inspect the soil.

Shown schematically in Figure 1, The VisCPT consists of two B&W miniature cameras, lenses and lighting systems with individual housing units.

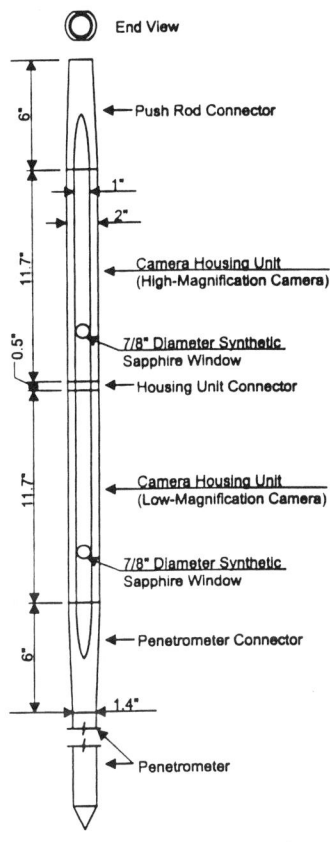

Figure 1. VisCPT Schematic
(Raschke and Hryciw, 1997)

The two cameras record the soil images through synthetic sapphire windows. Each camera system operates at a different level of magnification providing fields of view between 2 mm and 20 mm (diagonally). The images are recorded continuously in real time as the probe is advanced at the standard CPT rate of 2 cm/sec. A field-portable S-

VHS recording system and monitor allow for immediate field inspection as well as recording for later viewing and analysis. Details of the VisCPT may be found in Raschke and Hryciw (1997). Photographs of the VisCPT system and a stream of images may be found at the web site: www-personal.engin.umich.edu/~romanh/viscpt/ viscpt.htm.

After recording, the continuous stream of images is digitized at a rate of 30 frames per second using a Perception™ Video Digitizer Card installed on a microcomputer. The VisCPT creates large quantities of visual data. Digital images have a resolution of 720x480 pixels with a color depth of 256 gray-tone values. At a capture rate of 30 frames/second and a CPT advance rate of 2 cm/sec, 1500 images and approximately 460 MB worth of digital data are stored per meter of soil. As such, automated computer techniques are needed to archive and process the information. Digital image processing and analysis is performed utilizing Khoros®, a visual programming environment for software development, image and signal processing, machine vision and scientific visualization. Each image is digitized and analyzed by a set of image processing routines. In this study, detailed delineation of the soil stratigraphy from the digital soil profile was possible using statistical image textural analysis.

DIGITAL IMAGE TEXTURAL ANALYSIS FOR SOIL CHARACTERIZATION

A digital image is basically a two-dimensional grid of pixels or picture elements. During the digitization process, the brightness of the sampled scene is quantized and mapped to the corresponding pixel in the digital image. For gray-tone digital images, each pixel is assigned an integer value ranging from 0 (pure black) to 255 (pure white).

Image texture may be defined as "an attribute representing the spatial arrangements of the gray levels of the pixels in a region" (IEEE standard, 1990). The information contained in this spatial arrangement of gray-tones in the image is critical to the recognition and classification of these images. In this study, a statistical image textural analysis method known as the "Spatial Gray Level Dependence Method (SGLDM)", developed by Haralick et al. (1973) is utilized. This method quantifies 14 different textural indices based on the spatial distribution of the gray-tone pixels within the image. These indices relate to specific textural characteristics of the image such as homogeneity, contrast and the presence of organized structure within the image. The application of Haralick's indices to soil images was presented by Hryciw et al. (1998). A neural network was developed by Ghalib et al. (1998) to correlate 8 of the textural indices to mean grain size in uniform soils. The readers are encouraged to refer to these two papers for the analytical background to this work.

Variations in the texture of digitized soil images arise from several factors including: 1) grain size, 2) magnification, 3) illumination level, 4) darkening of zones around the soil grain boundaries affected by the lighting system, 5) color composition of the soil grains, and 6) natural variations of color and translucence within the surface of individual particles, especially when the particle footprint is a considerable portion of the total image size. It is important to note that the first two factors, grain size and magnification, are interrelated. Coarse sand at low magnification may possess a texture that is identical to that of a fine sand or silt at higher image magnification.

In order to demonstrate the capability of the textural analysis method in characterizing soil images, a layered soil profile was created in the lab from 5 different soils of varying average grain size and texture. The layering of the 78 cm long composite soil profile is shown in Figure 2(a) together with the respective soil description and minimum/maximum grain sizes. The sample was scanned by a black and white CCD video camera moving at a constant speed of 2 cm/sec, the advance rate of the standard CPT. Images were captured at a pixel resolution level of 63.1 pixels/mm. The stream of images was digitized in real time and archived for analysis. Examples of the 720x480 pixel images for each of the five soil types are shown in Figures 3(a)-3(e). The pixel per diameter (PPD) is defined as D_{50} measured in pixels.

In this study three of Haralick's textural indices, namely Energy (E), Contrast (CON), and Local Homogeneity (LH) were considered. Each index was found to possess a special differentiating feature that is important in the segmentation and classification process. The formal mathematical definitions of these indices are found in Ghailb et al., (1998).

Energy (E) is a measure of "global" homogeneity of the image. It is a measure of the agglomeration of large clusters of pixels with uniform gray-tones. In a homogeneous image, such as the clay image shown in Figure 3(e) there are very few gray-tone transitions. In fact, the whole image may be considered as one cluster with a narrow gray-tone band. This results in a large E value, as shown for the clay layers in Figure 2(b). As the particle diameter increases, the soil grains become more distinct and the image becomes more agglomerated forming small clusters of nearly uniform gray-tones. Furthermore, the variations in the gray-tones of individual clusters become more visible as shown in figures 3(a)-3(d). As a result, the global image homogeneity decreases with increasing grain size yielding lower values of E, as shown for the coarser-grained soils in Figure 2(b). Ghalib et al. (1998) demonstrated that as the grain size increases further and becomes a considerable part of the total image size (PPD>40), large clusters with nearly similar gray-tones form, yielding a relative increase in the global homogeneity of the image and again a larger E.

As shown in Figure 2(b) and the index summary in Table (1), the energy had two strong peaks corresponding to the locations of the two clay layers. Conversely, the E for all other soils was markedly lower. This distinguishing trend is only noticed in E, and thus it is considered to be a good index for detecting global uniformity in the image such as found in clay and silty clay soils.

Contrast (CON), is a measure of the local variation in pixel intensity in the image. CON is very sensitive to abrupt changes in local intensity generally located at boundary lines of adjacent particles with different gray-tones and at the boundaries of other

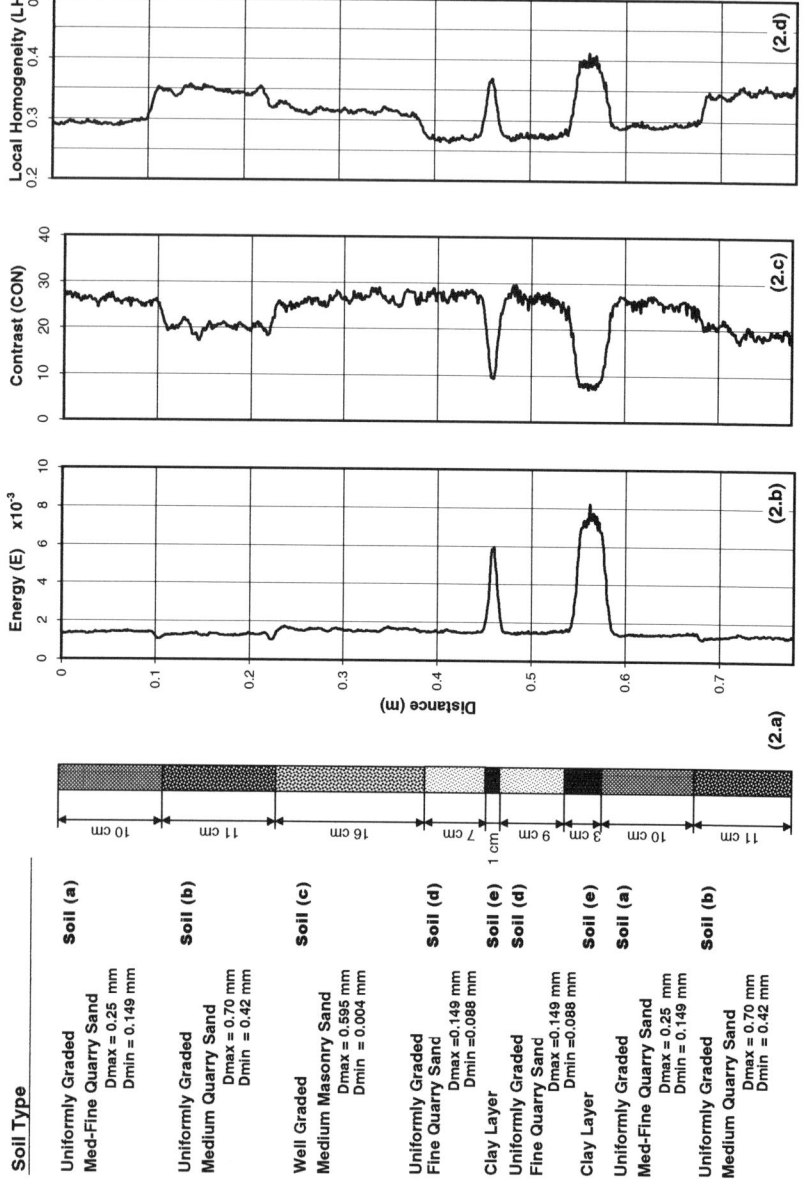

Figure 2. Soil Textural Analysis of Laboratory Layered Soil Profile

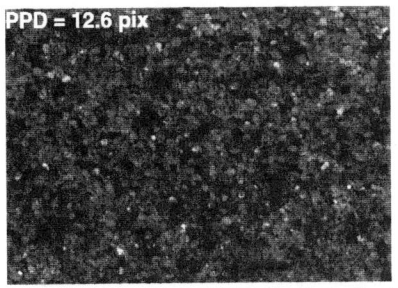

a. Soil (a): Uniform Graded Med-Fine Quarry Sand

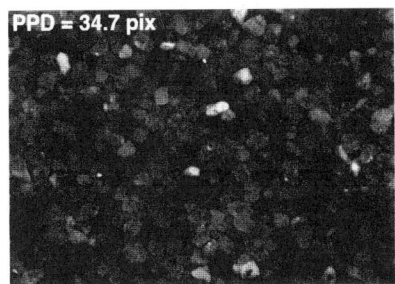

b. Soil (b): Uniform Graded Medium Quarry Sand

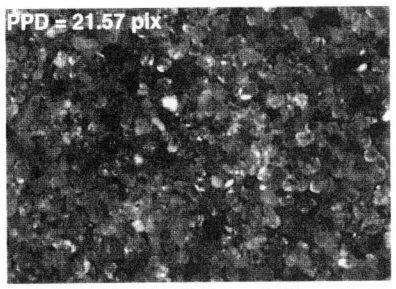

c. Soil (c): Well Graded Medium Masonry Sand

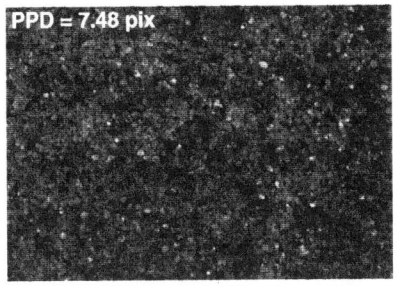

d. Soil (d): Uniform Graded Fine Quarry Sand

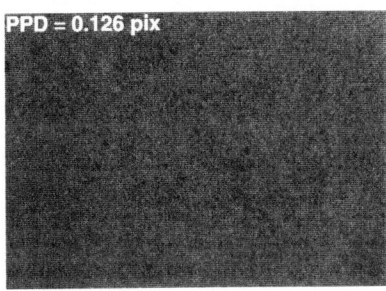

e. Soil (e): Clay

Figure 3. Sample Image of the 5 Soils Composing the Laboratory Soil Profile Specimen.

anomalies within the image such as fissures and open cracks. Naturally, when the PPD is below 1.0 as in Figure 3(e), the visual information contained in the grain and its boundary is integrated (blurred) in one pixel resulting in a uniform, non-contrasting image. This yields a low value for CON as clearly shown in Figure 2(b) for the clay. As PPD increases above approximately 5, particle edges become more distinct and the number of grain boundary pixels for a given image size increases. The increasing frequency of high contrast prone pixels will generally increase overall image contrast, as shown in Figures 3(a) & 3(d). On the other hand, for much larger PPD as in Figure 3(b), the frequency of the boundary pixels becomes smaller for the same image size yielding a somewhat lower CON, as shown for soil (b) in Figure 2(c).

As shown in Figure 2(c), CON is sensitive to the existence and the frequency of distinctive grains in the image. This index clearly differentiated sandy soils having different PPDs. Due to the lack of local gray-tone variation in the clay layer images, very low CON was observed. However, the ability to identify clay soils utilizing CON is not as strong as by E.

The CON for soils with extensive to moderate cracks is relatively high. As will be demonstrated in the next section, stiff desiccated soils with little overburden stress experience considerable macro cracking around the VisCPT camera housing due to the tensile stresses created by the lateral soil expansion and vertical shear created during VisCPT cone penetration. These cracks create localized zones of high textural contrast. CON is also sensitive to local anomalies in images of otherwise uniform soils. It will be demonstrated later that images of clayey or silty clay soils with local sandy lenses or seams generally yield high CON values.

Local Homogeneity (LH), is a measure of the degree of homogeneity at the "local" level. Images composed of a number of large locally homogeneous clusters, as typically found in coarse grained images with PPD>15 (Figure 3b), exhibit high LH. Also images of "global" homogeneity such as those for clay soils shown in Figure 3(e) have considerably higher LH. Conversely, the LH of soil images with small to medium clusters (3<PPD<15) such as the fine sand soil-(d) in Figure (3d) is considerably lower, as shown in Fig. 2(d) and Table (1).

Local homogeneity possesses the same differentiating characteristics as CON but with considerably lower noise. With reference to Table (1), the low values for the coefficient of variation for all soils compared to the two other indices indicate that LH is most suitable for differentiation and characterization. Also, LH indicated best correlation with particle D_{50} or PPD. As shown in Figure 2(d), the LH is consistent within different soil layers in the profile, has a sharp discontinuity at the exact location of the interface between soil layers and has a relatively low noise level compared to CON.

The above analysis and the results presented in Figure 2 demonstrate that the statistical textural indices correlated very well with PPD. Test results shown in Figure 2 and Table 1 indicate that the textural indices have excellent repeatability and differentiating capabilities.   In addition, the visualization and the camera speed in the laboratory setup, which simulated the prototype VisCPT faithfully characterized and delineated the soil profiles with layer thicknesses as small as 1 cm.

Table (1) Statistical Image Textural Analysis Summary (*)

Statistical Measurement		Soil Type and Size				
		Soil (a)	Soil (b)	Soil (c)	Soil (d)	Soil (e)
PPD	Average	12.6	34.7	21.57	7.48	0.126
Energy	Average x(10^{-3})	1.41	1.32	1.59	1.51	6.97
	Coefficient of Variation	0.037	0.049	0.046	0.036	0.100
Contrast	Average	25.8	19.9	26.4	26.9	8.3
	Coefficient of Variation	0.037	0.059	0.045	0.049	0.096
Local Homogeneity	Average	0.293	0.347	0.315	0.273	0.390
	Coefficient of Variation	0.012	0.016	0.017	0.015	0.034

(*) Images of boundary zones between two adjacent soil layers were excluded from the analysis.

IN-SITU SOIL CHARACTERIZATION AND THIN LAYER DELINEATION USING THE VisCPT.

A VisCPT soil penetration test was performed in a flood plain in Defiance County, Ohio. The objective of this study was to examine the capability of the image textural analysis algorithms in characterizing highly stratified soils and the ability to delineate thin features such as clay lenses and sand seams. A comparative study was made to evaluate the performance of both the standard CPT and the VisCPT in accurately characterizing the actual soil profile. Determination of the actual soil profile was made through high quality continuous soil sampling using a 6 inch long lined hand sampler. The sampling borehole was located within 0.6 m of the VisCPT test location. In the laboratory, the sample tubes were split in half longitudinally. One half was utilized for grain size analysis while the other half was utilized for layer delineation by visual inspection. The results of the sampling are shown in the far right column in Figure 4.

As indicated in the continuous soil profile shown in Figure 4, the 2.7 m deep soil profile consists of a top 1.15 m thick zone of interbedded silty sand and sandy silt layers followed by 0.9 m of stratified soil. This stratified zone had several soil layers with grain coarseness ranging from medium sand down to clayey silt. The individual soil layer thicknesses ranged from 15 cm to as thin as a few millimeters. The latter were commonly seams of clay interbedded between two coarser soils. Several distinctive thin clay and silty clay layers or lenses were found at 1.37 m, 1.90 m, 2.00 m and 2.03 m depths as shown in Figure 4. A more uniform soil zone of medium and fine sand

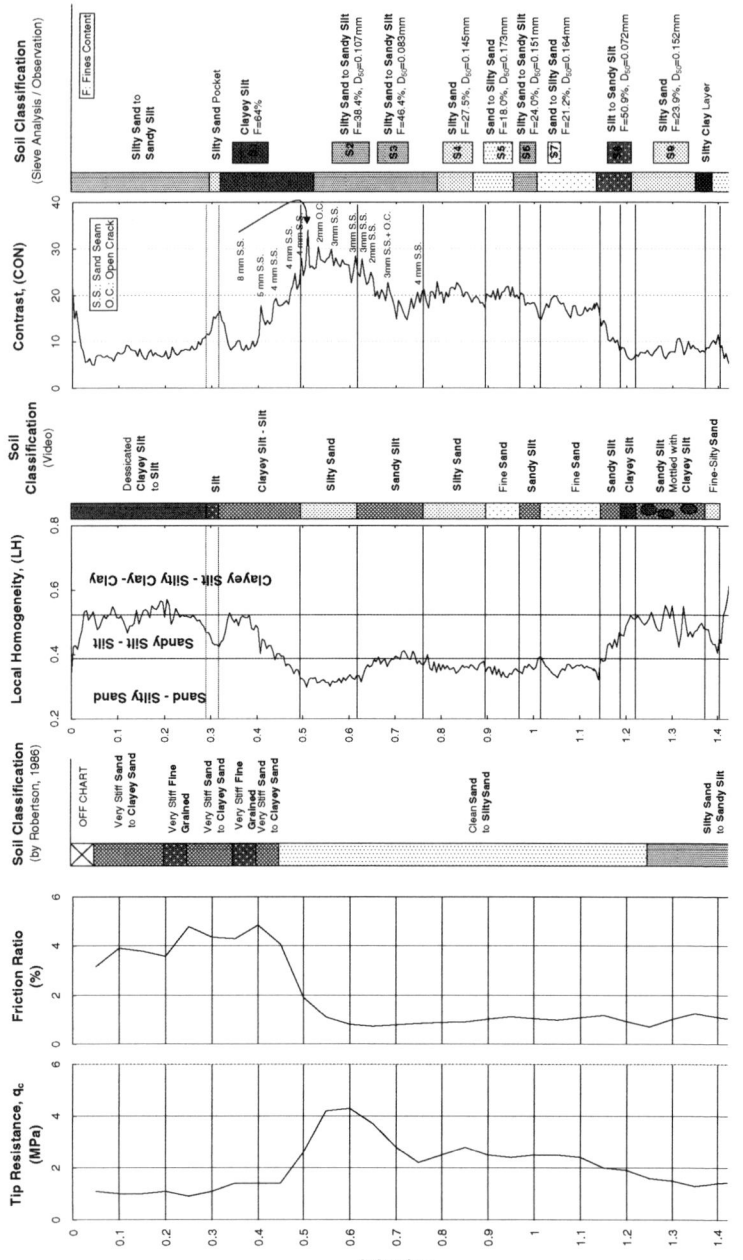

Figure 4. In-Situ CPT and Vis-CPT Image Texture Index Profiles.

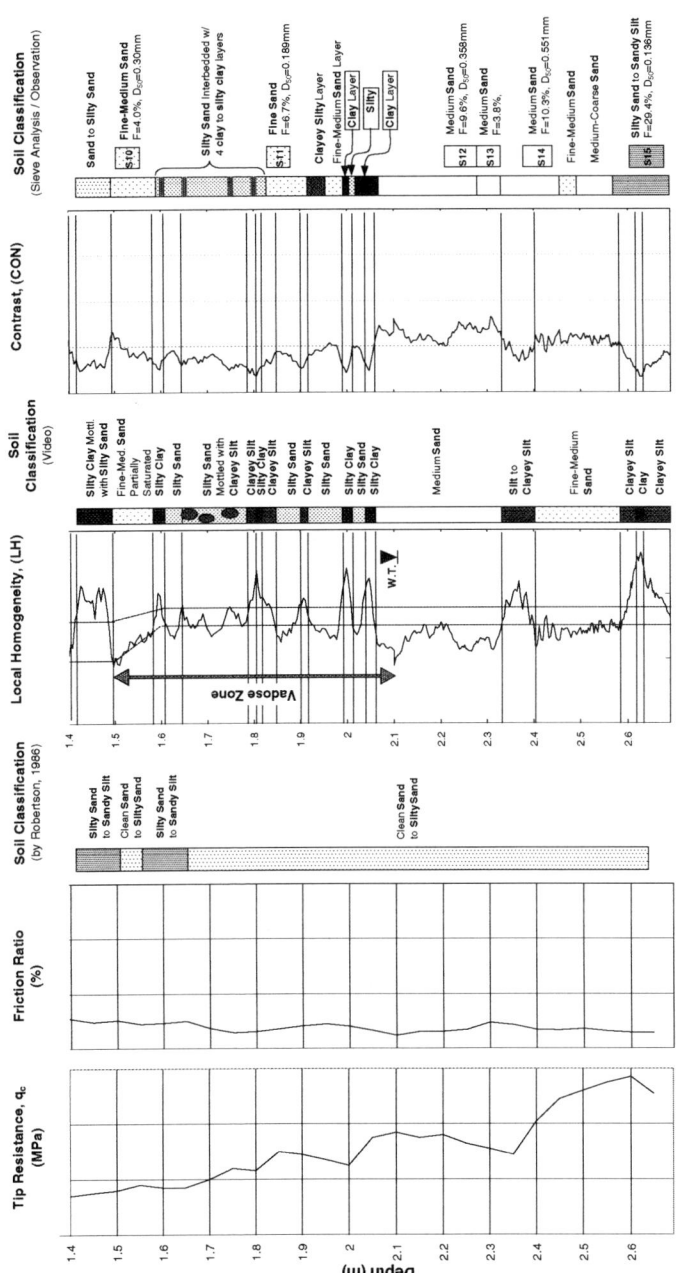

Figure 4. (cont.) In-Situ CPT and Vis-CPT Image Texture Index Profiles.

was encountered between 2.08 m to 2.58 m. Finally a silty sand layer was observed to the end of the borehole at 2.70 m.

A VisCPT test was performed and a continuous visual soil profile was recorded utilizing both high and low magnification cameras. Results from only the low magnification camera are presented in this study. The pixel resolution of the B&W camera is approximately 35 pixels/mm. Images were captured at a standard rate of 30 frames/second. The total recording time was approximately 128 seconds. The analog video signal was digitized to a series of standard 720x480 gray-tone digital images every $1/6^{th}$ of a second. This resulted in an approximate overlap of 50% between successive images. A total of 641 images were digitized with a total memory size just over 220MB. Since all VisCPT images are captured with a moving camera, images had to be de-interlaced utilizing a linear interpolation algorithm between the even scan lines of the original interlaced digital image.

Textural image analysis was performed for all images. The profiles of both LH and CON are shown in Figure 4. The soil classification column located between the two graphs was developed from direct observation of the digitized soil profile captured by the VisCPT. During the VisCPT test, both visual and standard CPT data are collected simultaneously. On the left side of Figure 4, the CPT tip resistance and friction ratios are shown. To the right side of the CPT data is the CPT-based soil classification according to Robertson (1990).

Generally, the textural analysis showed good agreement with the continuously sampled soil. CPT-based classification provided general agreement with the sampled soil profile, but at a greatly reduced level of detail.

Local homogeneity was able to capture the relatively minute variations between the silty sand and sandy silt layers located between 0.4 m and 1.15 m. Coarser sand layers are less homogeneous and thus yield a lower LH. The CPT-based soil classification predicted only one single homogeneous layer of clean sand to silty sand.

The relatively high amplitude peaks in the LH profile are associated with several thin silty clay layers between 1.15 m to 2.05 m. These silty clay layers ranged in thickness from 0.5 cm to 6 cm. The variation in the grain sizes of the alternating sand and silty clay layers of soil in that depth range is manifested by the corresponding stepped LH profile. Because of its high magnification, the VisCPT not only identifies thin anomalies, it can also show in detail the soil's internal structure. The silty clay layer located between depths 1.78 m and 1.85m in the LH profile is identified as a sequence of three layers with different grain coarseness, while the structure of the other double silty clay layers at depths of 1.98 m and 2.04 m are more distinct. This delineation is made possible by the high resolution of the VisCPT. It is noted that the desired resolution can be achieved by adjusting the magnification level of the VisCPT camera.

The exact depth and thickness of each layer predicted by the VisCPT varied slightly from the continuously sampled soil profile, as shown in Figure 4. This variation can be attributed to the local non-uniformity in the soil stratigraphy between the two sampling locations approximately 0.6 m apart. It is interesting to note that the CPT-based classification predicted a more uniform soil profile. None of the thin clay layers were clearly identified. Rather, the CPT tip resistance only showed a little softness associated with the clay layers located at depths of 1.8 m and 2.0 m. Similar local dips were also associated with the clayey silt layers at 2.35 m and 2.65 m depths.

A partially saturated zone was encountered at approximately 1.50 m. This can be visualized by the partly filled voids in the video images observed as round uniform gray bubbles. The average size of these bubbles was approximately the same as the average grain size of the soil.

The water table was located at approximately 2.1 m. The presence of water filling the voids increases image uniformity. As water fills in the space between the grains and the synthetic sapphire viewing window, it blocks the view of the soil grains not in direct contact with the viewing plane. In addition, soil grains in direct contact with the viewing plane show as small dark contact marks distributed in a uniform background image. Under such conditions, the textural indices are controlled by the density and spatial distribution of these contact points which is directly correlated to the soil grain sizes. This behavior unfortunately influences the values of LH for such soils. This phenomenon was noticed to affect only soils with clean fine to medium grained sands. Silty sands, silts and clayey soils generally have less identifiable voids and a considerably larger number of contact points. This phenomenon is clearly shown in the LH profile at the depth of the medium sand layers below the water line where the magnitude of LH for these layers correspond to that of sandy silt to silty soils. As such, an adjustment to the LH range of values for different soil types is necessary as shown in Figure 4.

The CON profile showed distinct peaks within otherwise uniform soil layers located between depths 0.40 m and 0.76 m. These peaks were associated with sand seams and open cracks in this zone. Due to space limitations, the energy profile is not presented. However, this profile showed localized peaks corresponding to locations of silty-clay and clay layers in the soil profile.

The texture index values are sensitive to a number of experimental parameters including: 1) lighting setup and illumination level, 2) camera magnification, 3) camera signal to noise ratio (SNR). As such, agreement between the textural index values of images of similar soils captured in the laboratory and in-situ have not been achieved to date. Furthermore, laboratory sandy soils are free of fines and dry. Moisture and fines in the in-situ soils generally yield more uniform images with higher LH, E and lower CON values. Future research is required to establish a more robust correlation between the proposed textural index values and the type of in-situ soils, with proper adjustment for moisture and fines content in the soil.

CONCLUSIONS

Laboratory and in-situ testing have demonstrated that the VisCPT can unambiguously delineate soil stratigraphy utilizing image texture analysis and other computer vision techniques. Soil layers as thin as 1 cm can be precisely delineated. The accuracy of the method has been demonstrated by comparison with high quality continuously sampled soil. Comparison with soil classification techniques based on the CPT revealed that the VisCPT can provide more detailed layer delineation in highly stratified soils. A CPTU test with a high frequency sampling rate would have provided a better layer delineation particularly below the ground water table. More work is required to calibrate in-situ image textural index values below the water table.

ACKNOWLEDGEMENT

The authors would like to thank Shin, Seung Cheol, a doctoral-candidate at the Civil and Environmental Engineering Department, University of Michigan, for his assistance during the in-situ testing and sampling phases of this study. Funding for this study was provided by NSF research grant CMS-9700128.

REFERENCES

Arango, I., (1997), "Historical and Continued Role of the Standard Penetration Test Method in Geotechnical Earthquake Engineering", *Third Seismic Short Course on Evaluation and Mitigation of Earthquake Induced Liquefaction Hazards*, San Francisco, CA. March.

Douglas, B.J., Olsen, R.S. (1981), "Soil Classification Using Electric Cone Penetrometer", *Cone Penetration Testing and Experience*, Norris, G.M. and Holtz, R.D. (eds.), ASCE National Convention. St. Louis, MO.

Ghalib, A.M., Hryciw, R.D. and Shin, S.C. (1998) "Image Texture Analysis and Neural Network for the Characterization of Uniform Soils," *Proc. ASCE Congress on Computing in Civil Engineering*, pp. 671-682.

Haralick R.M., Shanmugam K. and Dinstein I. (1973) "Textural Features for Image Classification", *IEEE Transactions on System, Man, and Cybernetics*, Vol. 3, No. 6, pp. 610-621.

Hryciw, R.D., Ghalib, A. M. and Raschke, S. A. (1998) "In-Situ Soil Characterization by VisCPT", *Proc. of the First International Conference on Site Characterization (ISC'98)*, pp. 1081-1086.

IEEE Standard 610.4-1990 (1990) "IEEE Standard Glossary of Image Processing and Pattern Recognition Terminology", *IEEE press*, New York, 1990

Lunne, T., Robertson, P.K., and Powell, J.J.M. (1997), "Cone Penetration Testing in Geotechnical Practice", *Blackie Academic and Professional*, London, pp. 312.

Mitchell, J.K. and Brandon, T.L. (1998), "Analysis and Use of CPT in Earthquake and Environmental Engineering", Geotechnical Site Characterization, Robertson and Mayne (eds), *Proceedings of the First International on Site Characterization-ISC'98*, Atlanta, Georgia, Vol. 1, pp. 69-97, April 1998.

Olsen, R.S. and Mitchell, J.K. (1995), "CPT Stress Normalization and Prediction of Soil Classification", *Proceedings, CPT'95*, Vol. 2., pp. 257-262, Linkoping, Sweden.

Raschke, S.A. and Hryciw, R. D. (1997) "Vision Cone Penetrometer (VisCPT) for Direct Subsurface Soil Observation", *Jour. of Geotechnical and Geoenvironmental Engineering of ASCE*, Vol. 123, No. 11, pp. 1074-1076.

Robertson, P.K., Campanella R.G., Gillespie, D. and Grieg, J., (1986), "Use of Piezometer Cone Data", Proceedings of In-Situ *'86 ASCE, Specialty Conference*, Blacksburg, Virginia, pp. 1263-1280.

Robertson, P.K. (1990), "Soil Classification Using the Cone Penetration Test*", Canadian Geotechnical Journal*. Vol. 27, pp. 151-158.

Robertson, P.K., Lunne, T. and Powell, J.J.M., (1998), "Geo-Environmental Applications of Penetrating Testing", Geotechnical Site Characterization, Robertson and Mayne (eds), *Proceedings of the First International Conference on Site Characterization, ISC'98*, Atlanta, Georgia, USA, April 19-22, Vol. 1, pp. 35-48.

Suzuki, Y., Tokimatsu, K., Koyamada, K., Taya, Y. and Kubota Y. (1995), "Field Correlation of Soil Liquefaction Based on CPT Data", *Proceedings, CPT'95*, Vol. 2, pp. 583-588, Linkoping, Sweden.

Treadwell, D. D. (1976), "The Influence of Gravity, Prestress, Compressibility, and Layering on Soil Resistance to Static Penetration", *Ph.D. Thesis*, University of California, Berkeley.

Vreugdenhil, R., Davis, R. and Berrill, J., (1994), "Interpretation of Cone Penetration Results in Multilayered Soils*", International Journal for Numerical and Analytical Methods in Geomechanics*, Vol. 18, No. 9, pp. 585-589.

MEASUREMENTS OF SIDE FRICTION
USING TEXTURED CPT FRICTION SLEEVES

By Jason T. DeJong,[1] S.M. ASCE, P. Ethan Cargill,[2] M. ASCE
and J. David Frost,[3] P.E., M. ASCE,

ABSTRACT

Currently, CPT data obtained from the friction sleeve measurement (f_s) is less widely used than the tip measurement (q_c). The "underuse" of the friction sleeve data is related to its high variability and the common sentiment that the sleeve measurement is unreliable. Recent research has quantitatively demonstrated the dominant influence of surface roughness on interface strength. Laboratory experiments have indicated that a change in the surface roughness from the conventional "smooth" configuration can more than double the interface friction. This effect is significant and has yet to be incorporated into geotechnical engineering practice.

A research program was undertaken to determine the effect of friction sleeve roughness on the measured sleeve resistance. A series of CPT soundings were performed using a set of roughened friction sleeves in addition to the conventional smooth friction sleeve. Results show the friction measurement to be heavily dependent on the surface roughness of the friction sleeve. The effect of surface roughness is material dependent, but in general the friction sleeve measurement obtained with a textured friction sleeve is about 1.6 to 2.0 times greater than a conventional "smooth" sleeve in granular soils. These results demonstrate the significant effect of surface roughness on f_s and the potential for modifying current in situ practice to account for this effect.

INTRODUCTION

The cone penetration test is becoming the predominant tool used for site characterization in the design of geotechnical systems. Its increased use has built on extensive research undertaken to develop correlations which allow the engineer to estimate the soil type, properties, and characteristics from continuous measurements.

---

[1] Research Assist., School of Civ. and Envir. Engrg., Georgia Inst. of Technol., Atlanta, GA 30332-0355.
[2] Design Eng., S&ME, Inc., 840 Low Country Boulevard, Mt. Pleasant, SC 29464.
[3] Prof., School of Civ. and Envir. Engrg., Georgia Inst. of Technol., Atlanta, GA 30332-0355.

Many of these correlations utilize both the friction sleeve and tip resistance measurements and more recent methods also incorporate the use of the pore pressure transducer data. Numerous direct design methods where parameters are determined directly from CPT data have been developed, in addition to the more conventional indirect methods, where CPT data are used to determine soil properties which are in turn used to estimate the required design parameters. Additionally, numerous other modifications have been made to the conventional CPT which enable measurements of additional insitu soil conditions including shear wave velocity, horizontal stress, resistivity, dielectric constant, pH, and pore fluid contamination (Lunne et al. 1997). With the widespread use of CPT data, the optimal performance of all components of the CPT device is of utmost importance.

The quality, reliability, and stability of the CPT measurements have been analyzed by a number of researchers including Lunne et al. (1986) and Tanaka (1996). Significant research into factors which affect the tip and pore pressure measurement has been performed. For the tip resistance measurement, the effect of temperature, wear, apex angle of the tip, and pore pressure have been understood and accounted for in design. Similarly, factors which affect the pore pressure measurement have been investigated and measures have been taken to account for these in design. Factors which have been identified to significantly affect the pore pressure measurement include the degree of saturation, pore pressure element location, axial load transmitted by the tip element, and soil type and density.

Similar to the tip resistance and pore pressure measurements, there are numerous factors which influence the friction sleeve measurement (Lunne et al. 1986, Tanaka 1996). Some of these factors have been previously addressed and accounted for in design, while others have been identified but no measures have been taken to account for or eliminate the effects. The CPT load cell arrangement can significantly limit the friction sleeve measurement resolution (Lunne et al. 1997). The subtraction cone load cell arrangement, where the friction sleeve measurement is determined by calculating the difference between the tip and total load, is least desirable as the resolution of the friction sleeve measurement is poor. CPT isolated load cell designs measure the tip and sleeve resistance independently and therefore significantly improve the resolution and reliability of the friction sleeve measurement. The resolution increases as the required full scale load cell capacity for the friction sleeve measurement decreases. Tanaka (1995) found the friction sleeve measurement to be 1/50 to 1/330 of the full scale capacity in clay and identified this as the primary cause of variability. The ASTM standard, D5778 (1995) estimates the standard deviation of the f_s measurement to be 15% of the full scale load cell output for the subtraction type designs while only 5% of the full scale load cell output for the isolated load cell designs. The importance of the CPT load cell arrangement has been recognized and isolated load cell arrangements are being incorporated into new CPT module designs.

The effect of sleeve wear and module stiffness on the friction sleeve measurement have also been researched and found to significantly affect the friction measurement. Jekel

(1988) showed that sleeve wear does occur during common use, potentially resulting in up to a 50% increase in the friction sleeve measurement. Recognizing the effect of wear, guidelines have been put forth relative to the control of the effect of CPT tip and friction sleeve wear through national (ASTM D-5778) and international (ISSMFE 1989; Swedish Geotechnical Society 1992) standards. Similarly, Vuong et al. (1988) showed through testing two different CPT modules that the stiffness of the CPT device can significantly affect the friction sleeve measurement due to an increased likelihood of bending. To minimize this effect, a full eight gage bridge transducer can be used and the stiffness of the device increased.

The distance of the friction sleeve behind the tip significantly influences the friction sleeve measurement as well. By varying the distance between the cone tip and the center of the sleeve in a set of field tests, Campanella and Robertson (1981) determined the effect of sleeve location to be most significant in dense sand. The f_s measurement was shown to increase from the conventional f_s measurement (with the sleeve positioned directly behind the tip) by about 45% when the sleeve is positioned about 20 cm behind the tip. It then decreases until it reaches a constant value (which is slightly lower than the conventional f_s value) when positioned 35 cm behind the tip. As discussed by Huntsman et al. (1986), this observed variation in f_s is primarily due to the variation in horizontal stress acting along a CPT which conceptually can be separated into two parts; the "baseline" horizontal stress equal to the natural insitu lateral stress prior to penetration and the variation from the "baseline" stress due to insertion of the CPT. This apparent variation in horizontal stress primarily occurs in a rapidly changing stress zone beginning around the CPT tip and extending part-way up the CPT shaft. Using cavity expansion theory, an increase of 2.2 to 5.8 times the "baseline" horizontal stress due to the stress zone is possible (Huntsman et al. 1986). In cohesionless soils, the variation in stress and size of the stress zone would be primarily a function of the initial void ratio, insitu stress regime, and the geometry of the penetrometer. Therefore, f_s measurement outside of this highly variable stress condition would be preferable and possible by positioning the friction sleeve up the CPT shaft instead of directly behind the CPT tip. Unfortunately, the effect of CPT friction sleeve location on the friction sleeve measurement has not been fully appreciated as all current designs position the friction sleeve directly behind the tip in accordance with national standards (ASTM D 5778-95).

Failure to account for the above factors in the design of friction sleeves has resulted in a more variable and less consistent measurement than possible. This is echoed by the common sentiment in practice that the friction sleeve measurement is less reliable and more variable than the tip and pore pressure measurements. It is also evident in the numerous correlations and design methods that discard f_s and only utilize q_c and u. However, even if the above factors were properly accounted for in CPT design, the friction sleeve measurement would likely remain highly variable and therefore "underused." Why is this? The authors propose that the current standard friction sleeve setup does not account for the effect of surface roughness on the friction measurement. Since the friction sleeve measurement is, in essence, an oriented axisymmetric interface shear test, research and understanding obtained from laboratory studies of soil-geomaterial interfaces can be transferred to the CPT friction sleeve measurement. The primary factor that significantly affects interface strength (in addition to normal load and soil density) is the relative surface roughness of the counterface material. The remainder of this paper focuses on this issue as it applies to the friction sleeve measurement. First, a framework for understanding the effect of surface roughness on the friction sleeve measurement is developed. This is followed by presentation and analysis of the results from field tests which corroborate the significant effect of surface roughness on the f_s measurement.

EFFECT OF SURFACE ROUGHNESS ON INTERFACE STRENGTH

Surface roughness has long been identified as having a significant influence on interface strength. However, due to limitations in experimental methods for quantifying surface roughness characteristics, the effect of surface roughness on interface strength was limited to qualitative analyses. Work performed by Potyondy (1961) concluded that the interface strength was influenced by four major factors; the moisture content, soil composition, surface roughness, and normal load. Similarly, Brumund and Leonards (1973) using a pull-out cylinder apparatus concluded that roughness significantly affected the interface friction angle, increasing it in value until it became equal to the sand friction angle, resulting in failure within the soil and not at the interface. Additionally, they concluded that the size, angularity, and surface texture of the individual sand grains also influenced the interface friction angle.

The use of quantitative measures of surface roughness in geotechnical engineering has been developed more recently, with Uesugi and Kishida (1986a, 1986b) and others making significant contributions. Using a simple interface shear apparatus, Uesugi and Kishida tested five granular materials against a set of steel plates, each with a different surface roughness. The steel surface roughness, average particle diameter, and sand mineralogy were found to strongly affect interface behavior while the specimen density was found to only affect the ultimate strength of the interface. Noting that steel surface roughness and average particle diameter are length scale measurements, they determined that surface roughness should be a relative parameter which describes how two different materials (i.e. the surface and soil particles) interact. To capture this, Uesugi and Kishida (1986b) proposed a modified roughness parameter, R_n, which is calculated by measuring the vertical relief between the highest peak and the lowest valley over a lateral distance equal to the average particle size (D_{50}) and dividing the value by D_{50}. As evident in Figure 1, their tests firmly established the dominant influence of roughness on interface strength. With all other parameters held constant and only the surface roughness increased, the interface strength continues to increase until it becomes equal to the soil resistance, at which time the failure zone moves away from the interface and into the soil. The maximum shearing resistance possible is thus equal to the soil shearing resistance. This relationship between surface roughness and interface strength has been verified and shown to exist for a number of other sand-geomaterial interfaces including fiber reinforced polymers, machined surfaces, and textured and smooth geomembranes (Dove et al. 1997; Frost and Han 1999). To quantify the roughness of these materials which varied significantly in relief, Dove and Frost (1996) implemented the Optical Profile Microscopy (OPM) method, a digital imaging technique. To quantify the roughness of these counter-face materials, a three dimensional surface roughness parameter, R_s, which is defined as the ratio of the actual surface area over the nominal surface area, was used. Dove et al. (1997) showed that surface roughness has a similar

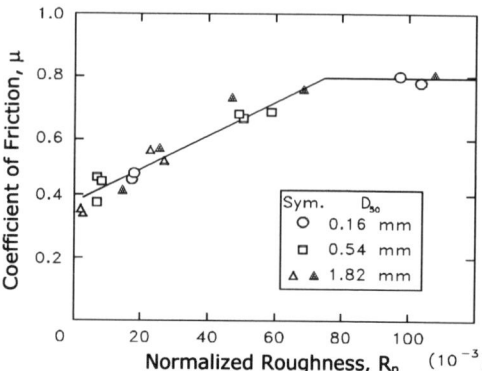

Figure 1. Surface Roughness vs. Friction Coefficient for Sand-Steel Interfaces (after Uesugi and Kishida 1986b)

effect on interface strength for sand-geomembrane interfaces and confirmed that increasing the interface roughness beyond a certain value does not increase the interface strength.

EFFECT OF INTERFACE MECHANISM ON THE FRICTION SLEEVE MEASUREMENT

The influence of surface roughness on the CPT friction sleeve measurement has not been fully appreciated to date. The ASTM D-3441 (1994) standard for the CPT specifies that the friction sleeve roughness, R_a (average roughness), must be equal to $0.50\mu m \pm 0.25\mu m$ (Figure 2). A series of surface roughness measurements made parallel to the CPT axis was performed on 10 and 15 cm^2 CPT friction sleeves from different manufacturers. Measurements were performed on new sleeves following shipment from the manufacturer and at different stages throughout their service life. The friction sleeve surface roughness (measured) for the new sleeves ranged from 0.28 to 2.08 μm, varying by manufacture and measurement location on the sleeve. Before use in the field, some of the new sleeves already exceeded the ASTM regulations. Through monitoring sleeve use in the field, the surface roughness was observed to change continuously, depending on the soil type encountered during the most recent penetration. The roughness of the used smooth friction sleeves ranged from 0.18μm to 6.85 μm, with significant changes observed after a change in site locations which corresponded to a change in soil type and/or stratigraphy. A further indication of continuous wear is the observed decrease or increase of surface roughness, which depended on the difference between the soil types encountered during consecutive soundings. These observations indicate that the interface conditions of the current friction sleeve measurement are not constant. The potential effect of the observed sleeve roughness variations are highlighted using the schematic shown in Figure 3.

Consider the interface friction angle – surface roughness relationships of two soils, one a clean medium sand and the other a silty sand, for a specific normal load and relative density. For the silty sand, the maximum interface friction angle (which is equal to the internal soil friction angle) will be reached at a relatively small surface roughness while the medium sand will reach its maximum interface friction angle at a larger surface roughness. More significantly,

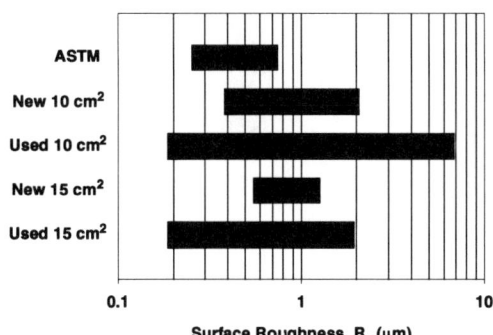

Figure 2. Surface Roughness Measurements of Conventional CPT Sleeves

the rate of change of the interface friction as a function of surface roughness is higher for the silty sand, with a slight change in surface roughness resulting in significant change in the interface strength. For the medium sand, the rate of change over this region is much lower, with a change in interface friction angle being negligible with a small change in surface roughness. It is noted that the same characteristic trend exists for all combinations of densities and normal loads, but with the maximum interface strength and the surface roughness at which the maximum interface strength is reached varying.

The effect of surface roughness on the CPT friction sleeve measurement is evident in Figure 3 as well. The interface strength mobilized when the silty sand is sheared against a new (smooth) friction sleeve is significantly less than when sheared against a used (roughened) friction sleeve, a result of high sensitivity to a small change in surface roughness. The same effect is present in interface strength mobilized when the clean sand is sheared against both new and used sleeves but to a lesser extent. The small change in surface roughness due to normal operating conditions results in different changes in mobilized interface strength, a condition undesirable for a standardized measurement. However, if the sleeve surface roughness was increased to where measurements in both soils occurred in their respective stable zones, this effect would be eliminated. In addition, the friction sleeve values would be measurements of the respective internal soil strengths.

TESTING PROGRAM

To evaluate the effect of surface roughness on the friction sleeve measurement, a research program involving CPT field testing was undertaken (Cargill 1999). Not withstanding the aforementioned concerns regarding the subtraction type load cell arrangement and the location of the friction sleeve, data for this initial investigation was

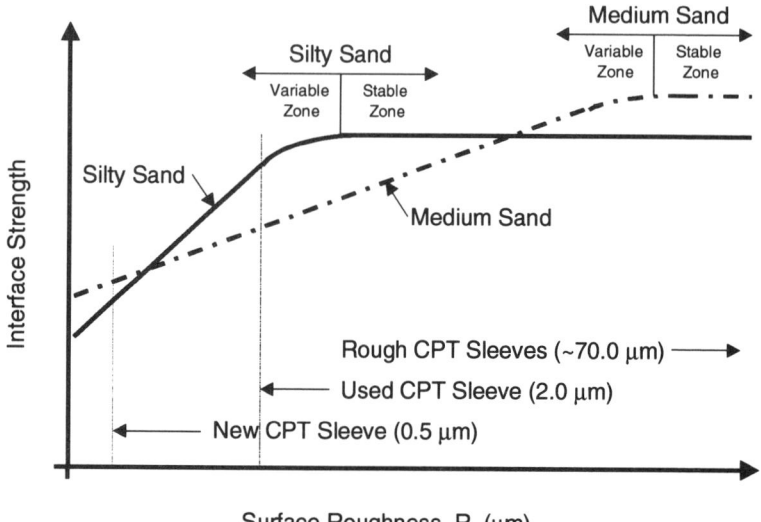

Figure 3. Schematic Illustrating the Effect of Surface Roughness on Interface Strength

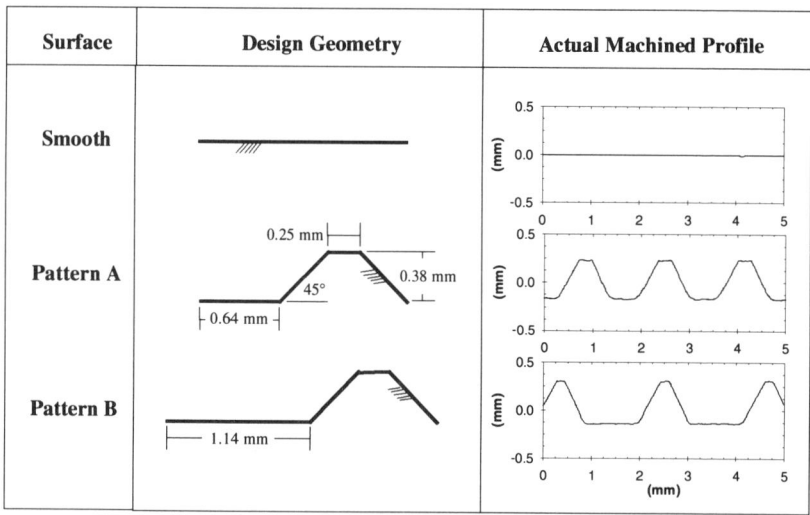

Figure 4. Plot of Smooth and Ribbed Surface Geometries and Profiles

obtained using a 15 cm^2 subtraction type CPT. A smooth surface, meeting the guidelines of the ASTM CPT standard, was used as a control surface to measure the behavior of the conventional friction sleeve. The two rough surface geometries (Pattern A and B) were selected which contained the same rib geometry, but varied in spacing between the ribs (Figure 4). A set of friction sleeves and a corresponding set of plates were machined with the specified textures, allowing for field and laboratory testing of the same surfaces. A stylus profilometer was used to non-destructively measure and monitor the change in surface roughness of the friction sleeves in between soundings. CPT soundings were performed at a penetration rate of 2 cm/s and a sampling interval of 5 cm in accordance with ASTM D 5778 (1995), with the exception of the friction sleeve surface roughness requirements.

To determine the effect of the textured friction sleeve on the sleeve friction measurement, a series of soundings were performed at a number of locations across the southeast United States. To verify the ability to compare adjacent profiles, tests were performed within 5 m of each other, with the minimum spacing governed by site conditions and truck reaction-anchor spacing requirements. Research by Keaveny et al. (1989) found the horizontal autocorrelation distance for q_c in sand to range from 14 to 38 m. Similarly, Phoon and Kulhawy (1996) found the scale of fluctuation for q_c in sand to range from 3 to 80 m, with a mean value of 47.9 m. To date, the horizontal autocorrelation distance in sand for the friction sleeve measurement has not been quantified. It was concluded that the effect of natural variability would be minimal if all soundings were performed within 5 m of each other. This was verified by comparing a set of profiles obtained with a conventional CPT sleeve. Results from one of the sites are shown herein.

Dredge Material Containment Area 2A (Savannah, Georgia)

Dredge spoil from Savannah Harbor and its approach are deposited in containment areas enclosed by a network of sand dikes. The horseshoe-shaped DMC 2A dike is located across the Savannah River from the turning basin of the Savannah Harbor and was constructed by dozing sand in lifts from the center of the containment area. Based on the construction methods and historical knowledge of the surrounding area, the dike is composed primarily of sand and the soil exposed on the sides of the dike can be considered representative of the soil contained in the dike. Five soundings were performed at DMC 2A to a depth of 10 m; one sounding was performed with the conventional smooth friction sleeve (SM) and two with each machined roughened friction sleeve (A1, A2, B1, B2). The water table is located below 10 m depth and therefore was not encountered during any of the soundings.

Soil stratigraphy determined using Robertson and Campanella (1983) based on the SM sounding (Figure 5) verifies that clean sand primarily exists to a depth of 10 m with thin silty sand layers exposed at the surface and between 7 and 8 m depth. Using the method proposed by Mayne (1995), the effective horizontal stress of the primary sand layer varied with depth and was estimated to reach about 78 kPa. The average relative density was calculated to be about 80% (Mayne and Kulhawy 1991) and also varied throughout the layer. Index properties of a sand sample from the site are summarized in Table 1.

Table 1. Summary of Soil Index Properties from DMC 2A

Site	G_s	e_{min}	e_{max}	D_{50} (mm)	C_u	% Fines
DMC 2A, Savannah, GA	2.65	0.56	0.89	0.40	2.8	≈ 1.3%

Tip resistance data for the five soundings were found to be similar, with an average tip resistance ratio of 1.14 and 1.23 for the A/SM and B/SM (Pattern A, B) ratios, respectively. The average ratios were calculated by first determining the ratio of the textured data to the smooth data at each depth increment for every textured profile. Ratios of profiles for the same textured sleeves were then averaged. The increase in the q_c measurements may be due to the textured friction sleeve altering the highly stressed zone around the CPT tip. Current research is investigating this potential interaction.

Figure 6 shows the friction sleeve measurement profiles obtained for the four roughened profiles (A1, A2, B1, B2), with the SM friction sleeve profile plotted as a reference. It is apparent that the textured friction sleeve profiles (A1, A2, B1, B2) are consistently higher than the smooth friction sleeve profile throughout the depth of the sounding. To

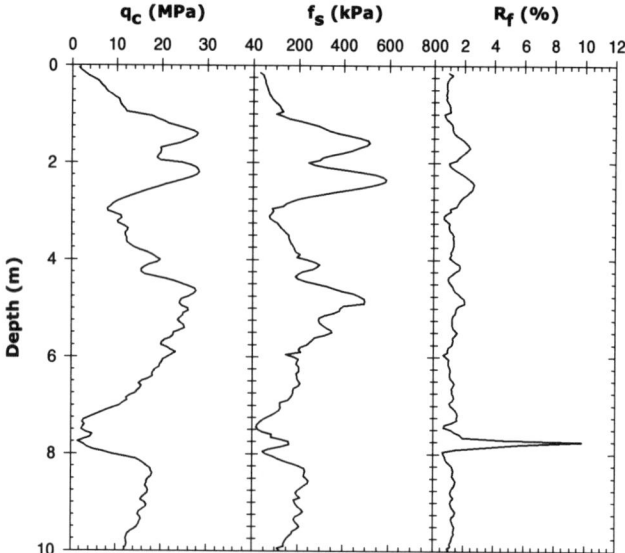

Figure 5. CPT Soundings with Smooth Friction Sleeve at DMC 2A.

quantify this difference, the ratios of the textured to smooth friction sleeve measurements (RSR) were calculated for each increment. On average, the A/SM and B/SM RSR values are equal to 1.88 and 1.80, respectively. In Figure 6 it is also evident that the ratio is not constant throughout the profile but varies, with RSR ratios at certain depths being approximately equal to 1.0 and greater than 3.0 at other locations. Further insight as to how the RSR ratios vary with depth can be gained by analyzing the profiles from the two types of textured friction sleeves in one meter increments (Table 2).

The incremental mean RSR values for the pattern A profiles range from 1.47 to 2.47. The effect of the textured surface on the friction sleeve measurement is clearly evident. The same is true for the B/SM one meter increment RSR values, with the mean values ranging from 1.33 to 2.54. Though this range is slightly larger than the range of average A/SM RSR values, the same trends exist.

Some one meter increments contain a larger range of RSR values than the majority of the increments, even though the mean RSR values are consistent with the other incremental means. For example, a larger range of RSR values is observed in the 2-3 and 3-4 m depth increment for the A/SM data. While the A1, A2, and SM profiles all contain similar trends, the peak in the SM profile at a depth of 2.30 m occurs at a slightly shallower depth than in the A1 and A2 profiles. Though this offset is only 15 cm and is likely a result of an uneven compacted lift of sand caused by the dozing lift method of construction, it influences the computed RSR values significantly. An offset is also observed in the B1 and B2 profiles, with B1 peak occurring above and the B2

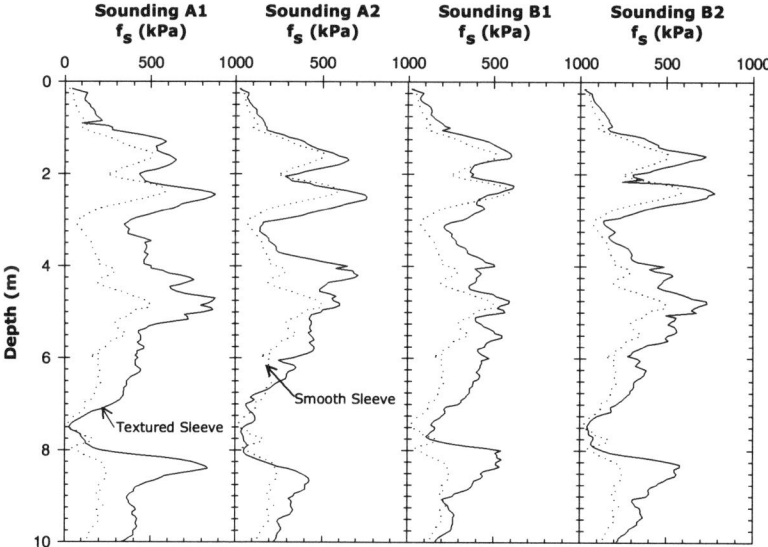

Figure 6. Rough Friction Sleeve CPT Profiles at DMC 2A.

peak occurring below the SM peak located at 2.30 m. Again, this offset influences the RSR values, increasing the range of values in the vicinity of the peak.

The 7-8 m increment also contains a greater range of RSR values spread about the mean. This depth coincides with a silty sand layer. In this increment, the maximum measured friction sleeve value is less than 250 kPa, which is less than that obtained throughout depths greater than 1 m. Since RSR is a relative measure of the difference between the rough and smooth friction values, it is sensitive to small changes in one or both of the friction sleeve measurements when the sleeve friction values are low and not very sensitive to small changes when the friction values are high. In this case, with the maximum measured friction sleeve value less than 250 kPa, the ratio is sensitive to small changes in the measured friction sleeve values. Additionally, friction sleeve values less than 250 kPa are approaching the limiting resolution for 15 cm^2 CPT used in this investigation. The depth increment from 8-9 m contains increases in the friction sleeve measurement in both the smooth (SM) and rough (A1, A2, B1, B2) sleeve soundings. The observed increase in the rough friction sleeve measurements are attributed to the transition from the silty sand layer into the bottom sand layer. In this transition, the space between the texture ribs which are initially partially clogged with the fine grains from the top layer are replaced by the sand. The additional force required to displace the fine grained soil may provide a sleeve friction increase for the rough sleeve measurements, a mechanism which does not occur for the smooth sleeve measurements.

Table 2. Average RSR Values for each One Meter Increment.

Depth Increment (m)	RSR Values	
	A / SM	B / SM
0.0 - 1.0	1.75	1.51
1.0 - 2.0	1.47	1.33
2.0 - 3.0	2.03	1.60
3.0 - 4.0	2.47	1.96
4.0 - 5.0	2.12	1.61
5.0 - 6.0	1.73	1.69
6.0 - 7.0	1.59	1.87
7.0 - 8.0	1.52	2.54
8.0 - 9.0	2.13	2.29
9.0 - 10.0	1.94	1.53
Cumulative Average	**1.88**	**1.80**

Note: A = f_s measurements with pattern A texture CPT sleeve
        B = f_s measurements with pattern B texture CPT sleeve
        SM = f_s measurements with conventional smooth CPT sleeve

CONCLUSIONS

A framework explaining the factors and mechanisms which influence the friction sleeve measurement were reviewed in depth. The load cell design (subtraction versus isolated), friction sleeve location, wear, and surface roughness were reported as factors which have not been fully appreciated and accounted for in CPT design. Previous research has clearly shown the adverse effects of the first three factors, while the effect of surface roughness has not been investigated thoroughly.

The significant effect of surface roughness on the friction sleeve measurement has been established herein, first through an understanding of the interface mechanism and later through CPT field tests. Previous work on factors which effect the interface mechanism clearly identify relative surface roughness to have a dominant effect on interface strength. The textured friction sleeve profiles were consistently higher than the conventional smooth friction sleeve profiles, with an average 80% increase in the sleeve friction.

The mechanisms governing the observed variability in the tests presented herein extend directly to all CPT testing performed in accordance with standardized guidelines. Therefore all f_s CPT data is affected by the surface roughness of the friction sleeve. The f_s profiles contain potentially significant variability in the measurement which is not reflective of the properties of the soil in contact with the friction sleeve, but rather variability due to the measurement system. This variability extends through all applications in which the f_s data is used in correlations for soil characterization and properties.

ACKNOWLEDGEMENTS

The work performed in this study has been supported by the National Science Foundation grant numbers #CMS-9457549 and #CMS-9700186. This support is gratefully acknowledged.

REFERENCES

ASTM. (1994). "Standard test method for deep, quasi-static, cone and friction-cone penetration tests of soil." D 3441-94, Philadelphia, PA.

ASTM. (1995). "Standard test method for performing electronic friction cone and piezocone penetration testing of soils." D 5778-95, Philadelphia, PA.

Brumund, W.F. & Leonards, G.A. (1973). "Experimental study of static and dynamic friction between sand and typical construction materials." *Journal of Testing and Evaluation*, 1(2), 162-165.

Campanella, R.G. and Robertson, P.K. (1981). "Applied cone research." *Proc., Symposium on Cone Penetration Testing and Experience: Geotechnical Engineering Division, ASCE*, October 1981, 343-362.

Cargill, P.E. (1999). "The influence of friction sleeve roughness on cone penetration measurements." *Master's Thesis*, Georgia Institute of Technology, 140 pp.

Dove, J.E. and Frost, J.D., (1996). "A method for measuring geomembrane surface roughness." *Geosynthetics International*, 3 (3), 369-392.

Dove, J.E., Frost, J.D., Han, J. and Bachus, R.C. (1997) "The influence of geomembrane surface roughness on interface strength." *Proc., Geosynthetics '97, Long Beach , CA*, 2, 863-876.

Frost, J.D. and Han, J. (1999) "Behavior of interfaces between fiber-reinforced polymers and sands." *Journal of Geotechnical and Geoenvironmental Engineering*, 125 (8), 633-640.

Huntsman, S.R., Mitchell, J.K., Klejbuk, L.W., Shinde, S.B. (1986). "Lateral stress measurement during cone penetration." *Proc., ASCE Specialty Conf. In Situ '86: Use of In Situ Tests in Geotechnical Engineering*, Blacksburg, VA., 617-634.

ISSMFE. (1989). Appendix A: "International reference test procedure for cone penetration test (CPT)." Report of the ISSMFE Technical Committee on Penetration Testing of Soils – TC 16, with Reference to Test Procedures, Swedish Geotechnical Institute, Linkoping, Information, 7, 6-16.

Jekel, J.W.A. (1988). "Wear of the friction sleeve and its effect on the measured local friction." *Proc., Penetration Testing 1988, ISOPT-1*, 2, 805-808.

Keaveny, J.M., Nadim, F., and Lacasse, S. (1989). "Autocorrelation functions for offshore geotechnical data." *Proc.,5th International Conference on Structural Safety and Reliability*, 1, 263-270.

Lunne, T., Eidsmoen, T., Gillespie, D., and Howland, J.D. (1986). "Laboratory and field evaluation of cone penetrometers." *Proc., ASCE Specialty Conf. In Situ '86: Use of In Situ Tests in Geotechnical Engineering*, Blacksburg, VA., 714-729.

Lunne, T., Robertson, P.K., and Powell, J.J.M. *Cone Penetration Testing in Geotechnical Practice*. Blackie Academic & Professional, New York, 1997.

Mayne, P.W. and Kulhawy, F.H. (1991). *Chamber Calibration Testing, ISOCCT-1*, Elsevier, NY, 197-211, 257-264.

Mayne, P.W. (1995). "CPT determination of overconsolidation ratio and lateral stresses in clean quartz sands." *Proc., CPT '95*, 2, 215-220.

Phoon, K.K. and Kulhawy, F.H. (1996). "On quantifying inherent soil variability." *Proc., Uncertainty in the Geologic Environment: From Theory to Practice*, 1, 326-340.

Potyondy, J.G. (1961). "Skin friction between various soils and construction materials." *Geotechnique*, 11, 339-355.

Robertson, P.K. and Campanella, R.G. (1983). "Interpretation of cone penetrometer test: Part I: Sand." *Canadian Geotechnical Journal*, 20 (4), 718-33.

Swedish Geotechnical Society. (1992). "Recommended Standard for Cone Penetration Tests." *SGF Report 1:93 E*. Swedish Geotechnical Institute, 26 pp.

Tanaka, H. (1995). "National Report - the current state of CPT in Japan." *Proc., CPT '95*, 1, 115-124.

Uesugi, M. & Kishida, H. (1986a). "Influential factors of friction between steel and dry sands." *Soils and Foundations*, 26 (2), 29-42.

Uesugi, M. & Kishida, H. (1986b). "Frictional resistance at yield between dry sand and mild steel." *Soils and Foundations*, 26 (4), 139-149.

Vuong, B., Donald, I.B., and Parkin, A.K. (1988). "Some aspects of the design of a friction cone penetrometer." *Proc., Fifth Australia-New Zealand Conference on Geomechanics*, 198-201.

USE OF THE SHARP CONE TEST FOR IN SITU DETERMINATION OF UNDRAINED SHEAR STRENGTH OF CLAY

B. Ladanyi[1], F.ASCE, H. Longtin[2] and A. Ducharme[3]

ABSTRACT

A recently developed in-situ testing method, called "The Sharp Cone Test" consists in pushing a low-angle truncated cone with 2 degrees taper into a smaller diameter prebored pilot hole. As the cone descends, it causes a continuous enlargement of the pilot hole, which, with a proper instrumentation, can be translated into a relationship between radial pressure and radial (or shear) strain, similarly as in a pressuremeter test. A first, preboring version of the sharp cone with three lateral pressure sensors, was tested in the field five years ago, and yielded positive results (Ladanyi et al. 1995). The present paper describes the use in the field of an improved, self-boring version of this instrument with four total-pressure transducers mounted on its lateral surface at different distances from its lower end. The new probe is able to furnish continuously four points of the pressure-expansion curve, which can be translated into a stress-strain relationship, using conventional pressuremeter data processing procedure. During 1998, the new instrument was tested in a thick layer of saturated clay at a site near Montréal. A comparison of the results with those obtained at the same site by some other types of tests, such as self-boring pressuremeter test and static cone test, was encouraging. The new testing method represents in fact a continuous and automated version of the pressuremeter test.

INTRODUCTION

The knowledge of mechanical properties of earth materials is an essential condition for the design of structural elements transfering applied loads to soils and rocks. For determining the mechanical properties of soils, both laboratory and in-situ methods are presently being used. In the former, undisturbed soil samples are taken from borings at selected levels, and

---

[1] Professor Emeritus, Dept. Of Civil, Geol. & Mining Engrg., Ecole Polytechnique, CP6079, Succ. Centre-ville, Montreal, QC, H3C 3A7, Canada

[2] Graduate Student

[3] Senior Technician

are subjected to certain tests pertinent to the purpose at hand. The in-situ methods, in turn, do not require soil sampling, but the scope of mechanical properties they are able to furnish is very limited compared with the former. On the other hand, the advantage of the latter is their ability to furnish a continuous picture of the geotechnical profile of the site in real time.

Presently used in-situ testing methods

In the geotechnical field, the most frequently used in-situ tests are the cone penetration test (CPT), the pressuremeter test (PMT), the flat dilatometer test (DMT), and the vane shear test (VST). Each of these tests produces in the soil a certain very distinct type of deformation and failure conditions, which makes it possible, by means of an appropriate interpretation method (or statistical correlation) to deduce from the results certain deformation and strength characteristics of the soil. The main features of these four in-situ test methods can be outlined as follows.

- The cone penetration test (CPT) is a continuous soil profiling method, that furnishes information on the soil strength properties and pore pressure response, but it yields very few direct data on soil deformability.

- The pressuremeter test (PMT) furnishes a fairly complete stress-strain and strength information on the soil at the test level, but it is not a continuous soil profiling method. In addition, it requires a complex field testing equipment and operation, which is difficult to automatize.

- The flat dilatometer test (DMT) has shown to be a useful and practical method for general soil profiling. However, as its theoretical interpretation is still unclear, its success is mainly due to statistical correlations with the results of some other types of field tests.

- The vane shear test (VST) is a relatively simple method for determining in-situ direct shear resistance of clays, and is particularly useful in connection with slope stability problems in clays.

The sharp cone test

The test consists in pushing a low-angle truncated cone (Fig. 1) into a pre-drilled or self-drilled cylindrical pilot hole of a smaller diameter. The purpose of the test is to produce in the soil the expansion of a quasi-cylindrical cavity, similarly as in a pressuremeter test. The test has up to now been carried out in two different manners.

(a) Testing in a creep mode. This method, called SCT, uses an ordinary non-instrumented cone. The test is carried out by holding constant the axial load applied to the cone, and by recording the relationship between the cone penetration and the time. Such a test makes it possible to determine the creep properties of a geological material, such as frozen soil, ice, rocksalt, and other materials having distinct creep properties. A complete theory and practical application of such a test was presented in three previous papers (Ladanyi and

Talabard, 1989; Ladanyi and Sgaoula, 1992; Leite et al., 1993).

(b) Testing in a continuous penetration mode. This method, called the Instrumented Sharp Cone Test (ISCT), which is the subject of this paper, uses a low-angle truncated cone, which is able to record by a system of pressure transducers (PT-s), installed at several levels of its lateral surface, the resistance of the soil against the enlargement of a predrilled pilot hole, caused y the cone penetration. The test is made in a manner similar to CPT, by making the cone to penetrate into the pilot hole at a steady penetration rate. Using current PMT interpretation methods, certain mechanical properties of the material can be deduced from the recorded relationship between the vertical penetration of the cone, which is directly related to the enlargement of the pilot hole, and the lateral pressure acting on the cone at several selected cone levels, recorded by lateral pressure transducers (Fig. 1). The test is intended to furnish soil information similar to PMT, but in a continuous manner, similar to CPT.

INTERPRETATION OF THE INSTRUMENTED SHARP CONE TEST

Figure 1 shows schematically a low-angle cone with 3 total pressure transducers (PT-s), installed at 3 levels of its lateral surface. The half-angle of the cone (the taper) is α. When the cone is pushed downwards, it gradually enlarges the pilot hole from radius r to the radius R of the main borehole. Assuming that a pressure transducer PT_i is placed at a level x_i from the upper end (I-I) of the pilot hole, it will record a pressure p_p necessary to enlarge the hole from r to

$$r_i = r + x_i \tan\alpha \qquad [1]$$

Adopting for simplicity the Gibson and Anderson's (1961) strain measure, $\Delta V/V$, which, for a volume constant plane strain case approximately equals the engineering shear strain, γ, one gets from geometrical considerations

$$\left(\frac{\Delta v}{v} \right)_i = 1 - \left(1 + \frac{\Delta r_i}{r} \right)^{-2} = 1 - \left(\frac{r}{r_i} \right)^2 \qquad [2]$$

where V denotes the current volume of the expanded cylindrical hole, and ΔV its cumulative increase.

At any given lateral pressure measurement level, the strains will remain the same, as long as the pilot hole precedes the cone, but the recorded pressure will vary according to the soil properties. In the continuous penetration mode, taper angles of 1 to 2 degrees are found convenient for testing saturated clays, because they are able to cover the most important portion of the stress-strain curve. On the other hand, larger angles of up to 5 degrees and more may be found more appropriate when testing very compressible or weak materials.

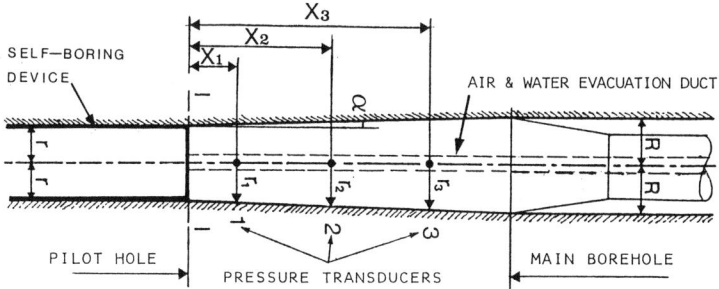

Fig. 1. Instrumented Sharp Cone (Schematic)

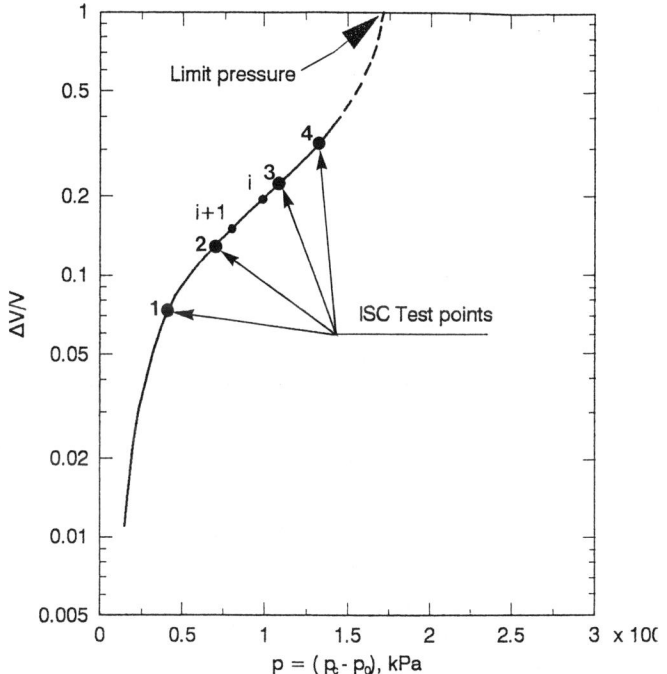

Fig. 2. Schematic Pressuremeter Curve in a Semi-log Plot with 4 Points that could be obtained by an Instrumented Sharp Cone Test.

Following Gibson and Anderson's (1961) original idea, Ladanyi (1972) has shown that the knowledge of the p_i versus ($\Delta V/V$) relationship makes it possible to determine the whole stress-strain curve of the soil in plane strain undrained shear, by applying the solution to any two consecutive points (i, i+1) on the pressuremeter curve (Fig. 2), from which the undrained cohesion under axial symmetry conditions, c_u, can be calculated

$$q_{ps} \equiv (\sigma_1 - \sigma_3)_{ps} = \frac{2 (p_i - p_{i+1})}{\ln (\Delta V/V)_i - \ln (\Delta V/V)_{i+1}} \qquad [3]$$

$$c_u = (\sqrt{3}/4) \, q_{ps} \qquad [4]$$

The corresponding axial symmetry compression strain, ϵ_{1a}, is related to the average shear strain, $\gamma_{i, i+1}$, by: $\epsilon_{1a} = \gamma_{i, i+1}/\sqrt{3} = 0.577 \, \gamma_{i, i+1}$.

In the case of an ISC Test, if lateral pressure transducers are installed at n levels, this will give n points of the "pressuremeter curve", or n points (n-1 plus the origin) of the resulting stress-strain curve. For the Instrumented Sharp Cone used in the present tests, where only 4 lateral PT-s were installed, only 4 points of the stress- strain curve could have been determined by this method.

TEST EQUIPMENT AND PERFORMANCE

The instrumented sharp cone with 2 degrees taper and with 4 lateral pressure measuring levels (Fig. 3), was designed by the first author and realized by Roctest Ltd of Montreal. It uses a set of lateral pressure measuring transducers (Sensotec, Model S/1542-03, Cap. 200 psig) (Fig. 3). The size of the cone and the position of the transducers is shown in Table 1. The $\Delta V/V$ values are seen to cover the most interesting interval of shear strains close to the failure of saturated clays.

The test procedure consists in first drilling the pilot hole of a selected diameter, which should remain stable until the start of the test. This is possible in a relatively stiff clays, but requires a self- boring device with the use of drilling mud in weak soils. The cone is then pushed into the pilot hole at a selected rate and the pressure variation with depth on the pressure transducers is recorded. The necessary equipment includes a drilling rig, a CPT-type penetration rig, a self-boring devicen and a data acquisition system (Fig. 4).

Fig. 3. Sharp Cone with 4 Levels of Lateral Pressure Transducers,

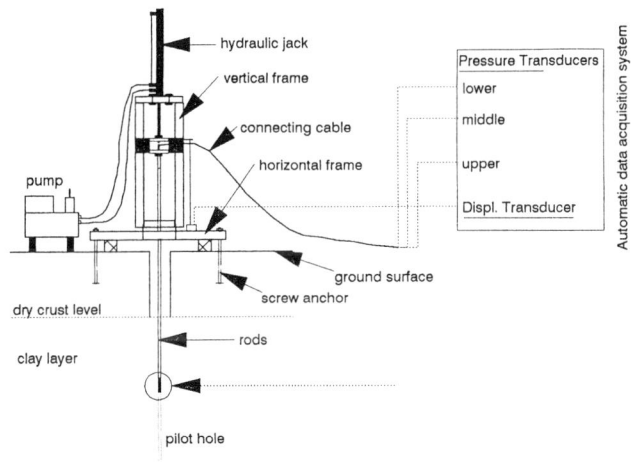

Fig. 4. Loading and Data Acquisition System.

TEST SITE

Three ISC tests (and two PMT tests, for comparison) were carried out in the Fall of 1998 at a test site near St-Hilaire, Québec, about 20 km east of Montréal. At the site, the soil profile consists of about 1.5 m thick dried crust, underlain by a thick layer of gray saturated NC clay. The layer in which the tests were made, extends from 2 m to about 5 m. The clay has a dry density of 1.53 Mg/m^3 , water content w = 70 - 90%, and the vane test sensitivity of about 17. Its Atterberg Limits are: w_p = 25% and w_L = 63 to 72%. The lateral at rest pressure coefficient K_o determined at the site in an earlier study, is about 0.45. The ground water level during the tests was about 1.2 m below the ground surface.

EQUIPMENT USED IN THE TESTS

A hole of 4" (10 cm) diam. was first drilled through the upper crust of the ground. The hole was subsequently enlarged to 10" (25 cm) and protected by a steel casing. The pilot hole was then formed continuously by means of a self-boring system of the type described in Benoit & Clough (1986), using a 135 cm long hydraulic jack, activated by a hydraulic pump. The jack was installed on a steel frame, anchored into the soil by two 80 cm long screw anchors (Fig. 4). The ISC, installed on the self-boring device, was pushed into the pilot hole at a rate of 6 to 7 cm/min by means of the same hydraulic jack system (Fig. 4). This penetration rate was selected to approximate the average strain rate produced by an ordinary pressuremeter test.

The recording and data acquisition system consisted of 4 pressure transducers, located on the lateral surface of the cone, and one displacement potentiometer with 1500 mm range, installed on the rig (Fig. 4). The data from the four transducers were automatically recorded against the time on a 21X Micrologger (Campbell Scientific Inc.), which can be programmed either directly, or through a personal computer (Toshiba T2000SX). The values of lateral pressures were deduced from recorded excitations by using transducer calibration data supplied by Roctest Ltd. The calibration was subsequently independently checked in the laboratory by putting the cone in a sealed steel tube and applying water pressures from 0 to 700 kPa.

TEST RESULTS

In 1998, three ISC tests were carried out at the St-Hilaire site in three separate boreholes, drilled at a distance of about 1.50 m from one another. The results of two of them are shown in Fig. 5 (a) and (b), together with the calculated total lateral pressure, p_o , and the hydrostatic pressure, u. The recorded lateral pressures at 4 values of lateral strain, are seen to increase steadily with depth below the dried upper crust, in which much higher resistances were recorded. Note that, because of high sensitivity of pressure transducers, the curves had to be smoothed-up, using Statistica '98 program.

The continuity of the recorded pressures makes it possible, in principle, to define at each level of the soil profile, 4 points of an equivalent "pressuremeter curve", corresponding to the 4 pressure transducers. Figure 6 (a) and (b) shows axial stress-strain curves deduced from four arbitrarily selected levels of the ISCT curves in Fig. 5. Because of a small number of recorded points, the curves came out fairly irregular, but the general trend to strength increase with depth is clear. However, the main advantage of this test is that any

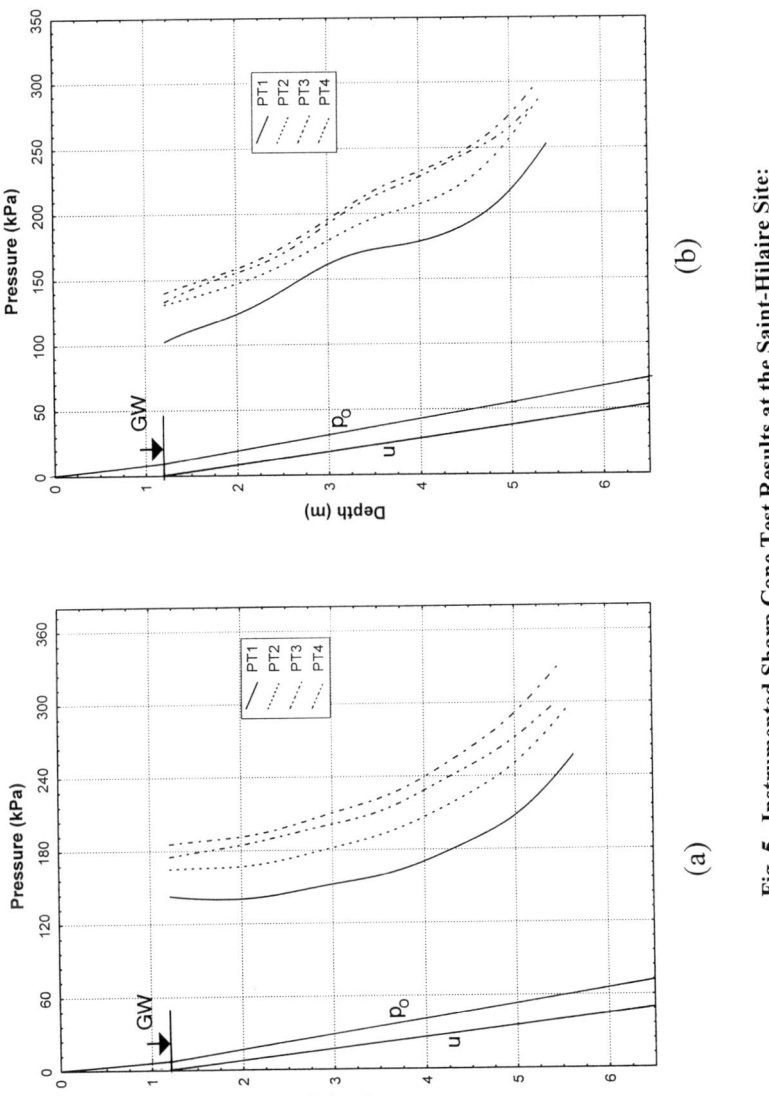

Fig. 5. Instrumented Sharp Cone Test Results at the Saint-Hilaire Site:
(a) September 1998, (b) October 1998.

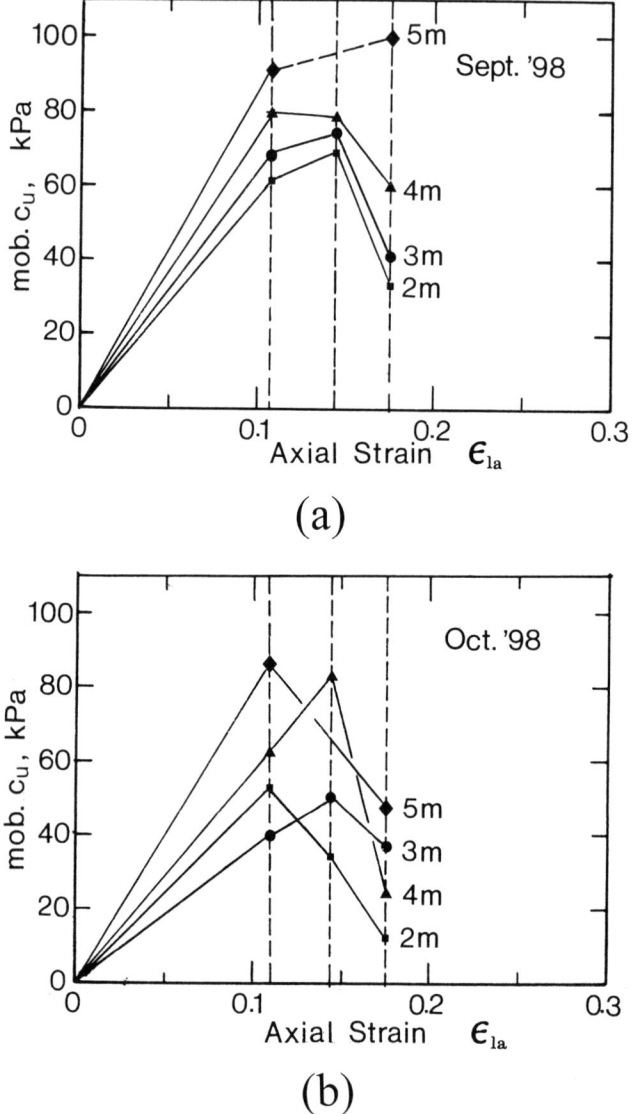

Fig 6. Axial Stress-Strain Curves Deduced from Four Arbitrarily
Selected Levels of ISC Tests (a) and (b) in Fig. 5.

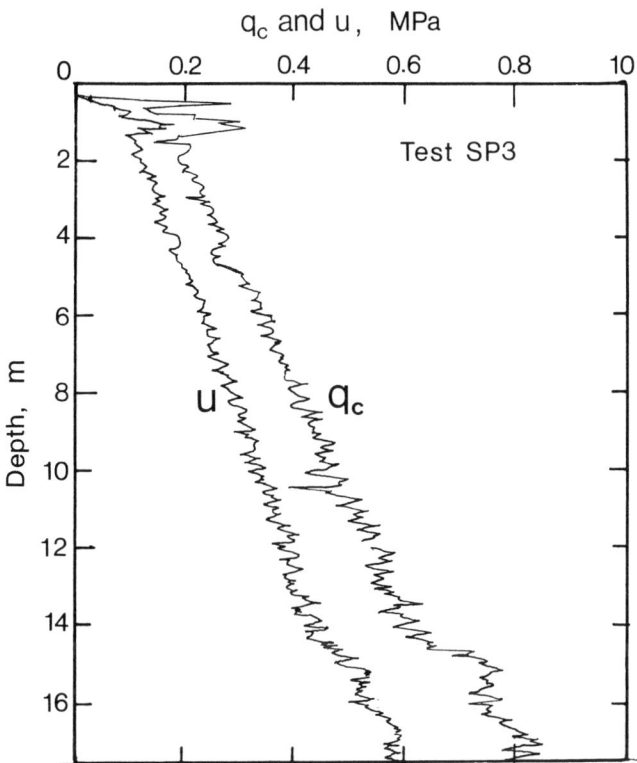

Fig. 7. Results of a Piezocone Test Obtained at the Same Site in 1987.

desired number of such curves can be generated from the tests at any selected number of levels. Clearly, with the addition of more lateral pressure transducers, smoother stress-strain curves could be obtained.

COMPARISON WITH OTHER TYPES OF TESTS CARRIED OUT AT THE SAME SITE

For comparison, two self-bored pressuremeter tests were carried out at a distance of about 2 m from the ISCT-S. Both PMT-s were made in the same borehole at two different levels, i.e., PMT1 at a depth of 2 m and PMT2 at 4 m. They gave plane strain undrained cohesion values of about 35 kPa, corresponding to the axial symmetry values of $c_u = 30$ kPa.

Earlier (Lafleur et al., 1987), several piezocone tests were also carried out at the same site. Figure 7 show one of the piezocone profiles down to 10 m depth. The presence of the upper crust is clearly seen in this profile, while the cone resistance, within the range of ISC Tests (2 to 5 m) is seen to vary between 200 and 300 kPa. Taking for this weak clay the N_c value of about 10, one gets c_u values between 20 to 30 kPa.

A comparison of these results with those obtained from ISC Tests in Fig. 6, shows that the latter give generally higher peak strengths, but in terms of an average strength they come closer to those recorded in the two above mentioned types of tests.

CONCLUSION

The field experience with the new version of the Instrumented Sharp Cone Test described in this paper, has given confidence that the new system represents an interesting and workable alternative to some presently used field tests.Compared to the pressuremeter test, the proposed method furnishes the data comparable to those measured by the PMT, but it can be used for a continuous sounding, which is not possible with the present PMT equipment. In addition, its capacity is not limited by the loading system as in a PMT, which makes it applicable also to very strong and stiff materials, such as weak rocks. Compared to the cone penetration test, the proposed method can be used for continuous sounding like CPT, but it furnishes more complete information on the stress-strain behavior of the soil. On the other hand, the proposed method requires a predrilled pilot hole, which must remain stable before and during cone penetration. This is easy to realize in stiff materials, but it may present certain difficulties in weak soils, in which a self-boring tool with the use of drilling mud will be necessary. The field study, using an upgraded cone with 4 lateral pressure transducers, described in this paper has given promising results. More such tests in soils of different type and stiffness are planned for the future.

ACKNOWLEDGEMENTS

Financial support from the Natural Sciences and Engineering Council of Canada is gratefully acknowledged. The detailed design and fabrication of the instrumented sharp cone was made by the Roctest Ltd, Montreal.

REFERENCES

Benoit, J. & Clough, G.W. (1986). Principal stresses derived from self-boring pressuremeter tests in soft clay. In: *"Pressuremeter and its Marine Applications"*, *2nd Int. Symp. on Pressuremeters* (Briaud & Audibert, Eds.), ASTM, STP 950, 137-149.

Gibson, R.E. and Anderson, W.F. (1961). In situ measurement of soil properties with the pressuremeter. *Civ.Eng. & Public Works Rev.*, London, May 1961, 615-618.

Ladanyi, B. (1972). In-situ determination of undrained stress-strain behavior of sensitive clays with the pressuremeter. *Canad. Geotech. J.*, 9, 313-319.

Ladanyi, B. (1994). Some unconventional field testing methods for earth materials. *Proc. Symp. on Developments in Geotech. Engrg.* 1936-1994, Bangkok, Thailand, 1-115-1-122.

Ladanyi, B. and Sgaoula, J. (1992). Sharp cone testing of creep properties of frozen sand. *Canad.Geotech. J.*, 29, 757-764.

Ladanyi, B. and Talabard, PH. (1989). Sharp cone testing of frozen soils and ice. *Proc. 5th Int.Conf. on Cold Regions Engrg.*, St.Paul, Minnesota, 282-296.

Ladanyi, B., Mchaileh, J. & Ducharme, A. (1995). A continuous pressuremeter test based on the "sharp cone" principle. In: *"The Pressuremeter and its New Avenues"*, *4th Int. Symp. on Pressuremeters*, (Ballivy Ed.), Sherbrooke, QC, 185-192.

Lafleur, J., Chiasson, P., Asselin, R. & Ducharme, A. (1987). L'évaluation des risques pour les travailleurs dans les excavations. *Rapport de recherche*, GEO-87-001, Dép de génie civil, École Polytechnique, Montréal

Leite, M.H., Ladanyi, B. and Gill, D.E. (1993). Determination of creep parameters of rocksalt by means of an in situ sharp cone test. *Int.J. of Rock Mech. and Mining Sciences*, 30, 3, 219-232.

TABLE 1. Data on the sharp cone used in the tests

Level	Description	$d_i = 2r_i$ cm	x_i cm	$(\Delta V/V)_i$
0	Top of pilot hole	7.35	0	0
1	PT1	7.97	8.90	0.15
2	PT2	8.33	14.00	0.22
3	PT3	8.65	19.10	0.28
4	PT4	9.00	24.20	0.33

AN ELECTRO-VIBROCONE FOR SITE-SPECIFIC EVALUATION OF SOIL LIQUEFACTION POTENTIAL

By Alec McGillivray[1], Thomas Casey[1], Paul W. Mayne[2], and James A. Schneider[3]

ABSTRACT

An electric downhole vibrocone has been designed for the site-specific evaluation of soil liquefaction susceptibility in high seismicity regions. The device induces localized cyclic pore pressures while concurrently measuring dynamic tip and friction resistances. Initial versions operated on a pneumatic impulse to generate dynamic forces and showed success in preliminary trials in historic liquefaction sites near Charleston, SC. The latest design couples a dual-element penetrometer with a piezo-actuator shaker and is completely electronic. This vibrocone offers a number of advantages over its predecessors including downhole vertical oscillation, adjustable dynamic displacement, control of excitation frequency from 0 to over 500 Hz, and the simultaneous measurement of porewater pressures at multiple positions including both midface and shoulder locations.

INTRODUCTION

The evaluation of soil deposits in high seismicity regions includes laboratory testing and field investigations to determine the likelihood of liquefaction, as well as the subsequent post-cyclic strength behavior. Current laboratory and field methods are fraught with empirical corrections and modifiers that result in high uncertainty for many sites, particularly the seismic zones of New Madrid/MO and Charleston/SC. In the laboratory approach, cyclic triaxial and cyclic simple shear tests on cohesionless materials rely often on reconstituted samples, that really do not contain facets of the

---

[1] Research Asst., Geosystems Program, Georgia Institute of Technology, Atlanta, GA 30332-0355.
[2] Assoc. Prof., School of Civil & Env. Engrg., Georgia Institute of Technology, Atlanta, GA 30332-0355; Email: pmayne@ce.gatech.edu
[3] Staff Engineer, Geosyntec Consultants, 1100 Lake Hearn Drive, Suite 200, Atlanta GA 30342-1523.

original soil fabric, inherent structure, void constituents, aging effects, cementation, and geostatic stress state. The obtaining of high-quality samples by freezing is very costly and not readily available on routine projects. Moreover, the utilization of cyclic laboratory tests relies on a number of empirical modifiers that attempt to account for initial stress state (K_o), overburden stress level (K_σ), depth effect (r_d), number of cycles to failure (N_f), and accumulated shear strains (Υ_s). The influences of time, sampling disturbance, fabric, and re-establishment of the in-situ anisotropic stresses are difficult or impossible to determine.

For the majority of projects, the greatest emphasis in assessing soil liquefaction potential is given to the results of field in-situ tests, primarily using the standard penetration test (SPT). Unfortunately, the number of corrections that must be made to the raw penetration data from even the most carefully conducted tests leads to great uncertainty. The SPT-N value must be modified to include corrections for energy efficiency, overburden stress level, fines content, borehole diameter, barrel liner, rod lengths, aging, overconsolidation ratio, and other factors (Skempton 1986; Kulhawy & Mayne 1990; Robertson & Wride 1997, 1998).

For liquefaction analysis, Seed et al. (1975) established a relationship between the adjusted $(N_1)_{60}$ and cyclic stress ratio (CSR = cyclic shear stress normalized to vertical effective stress). The CSR relates earthquake acceleration to the dynamic shear stress of the soil and can be obtained through a simplified procedure by Seed & Idriss (1971):

$$CSR = \frac{\tau_{cyc}}{\sigma_{vo}'} = \frac{\tau_{ave}}{\sigma_{vo}'} \approx 0.65 \frac{a_{max}}{g} \frac{\sigma_v}{\sigma_{vo}'} r_d \tag{1}$$

where τ_{cyc} = equivalent uniform cyclic shear stress, τ_{ave} = average cyclic shear stress, σ_{vo}' = effective vertical stress, σ_v = total vertical stress, a_{max} = peak ground acceleration, g = gravitational constant, and r_d = stress reduction coefficient accounting for the stiffness of the soil column. The energy-corrected SPT value, normalized to a stress level of one bar and designated $(N_1)_{60}$, is used in the well-known curves for a binary assessment of either "liquefaction" or "no liquefaction" (e.g. Glaser & Chung, 1995).

Recently, interest has focused on the development of CSR curves using the cone penetration test (CPT) to evaluate soil liquefaction potential (e.g., Suzuki, et al., 1995; Stark & Olson, 1995). The CPT offers several advantages over the SPT including better standardization, continuous records with depth, and the ability to measure several parameters, including: tip stress q_t, sleeve friction f_s, penetration porewater pressure u_b, Since the electrical readings are obtained downhole at the penetrometer level, there is not the need for the major correction of energy inefficiency, as with the older SPT resistance. However, normalization factors for overburden stress level and modifications for the effects of aging and fines content are still necessary (Robertson & Wride, 1998; Olson & Stark, 1998).

Other in-situ tests have been utilized for liquefaction assessment, including the flat plate dilatometer test (Reyna & Chameau, 1991) and shear wave velocity (Andrus & Stokoe, 1996). These two require normalization of the measured parameters for stress level and have been applied to more limited databases. The seismic cone penetration test (Campanella, 1994) is of particular interest in that it produces four separate measurements (q_t, f_s, u_b, and V_s), thus allowing two independent evaluations of liquefaction potential by comparing the CSR with both normalized tip stress (q_{c1}) and normalized shear wave velocity (V_{s1}). In addition, the downhole V_s is useful in assessing ground shaking and site amplification (e.g., SHAKE) for the direct evaluation of CSR at specific sites.

Each of the above approaches provides only an empirical and indirect assessment of soil liquefaction potential with uncertainty in the normalization and corrections. Therefore, a more rational and direct method was sought in the form of a vibrocone penetration test (VCPT).

VIBROCONE CONCEPT

The basic concept of a vibrocone consists of a cone penetrometer with trailing vibrator unit, as shown by Figure 1. Prior versions of a vibrocone device were developed in Japan, Italy, and Canada, and a detailed review is given by Wise, et al. (1999). The original model of the Japanese vibrocone was built by the Public Works Research Institute (PWRI) and applied a downhole horizontal centrifugal force of 32 kgf at a frequency of 200 Hz (Sasaki & Koga, 1982). Each test required two sister soundings [one static and one dynamic] with results compared by superposition of the q_c records. This was followed by later designs with dynamic forces of 80 and 160 kgf at 200 Hz (Teparaksa, 1987). A similar model developed in Italy also used a horizontal vibration source to induce localized liquefaction around the probe (Piccoli, 1993).

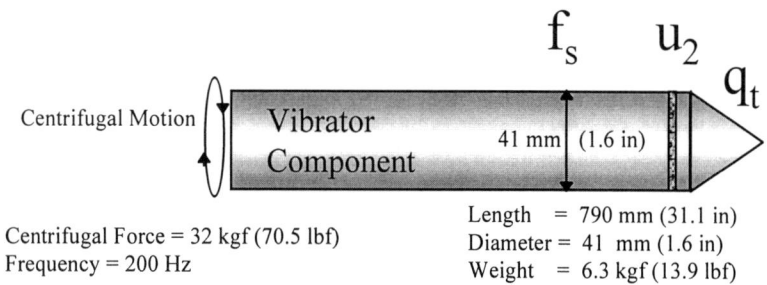

Figure 1. Schematic of the Original Vibratory Piezocone (Sasaki & Koga, 1982)

Previous field studies used simple side-by-side comparisons of static tip resistance (q_{cs}) with the dynamic tip resistance (q_{cd}). Figure 2 illustrates the recorded profiles of q_{cs} and q_{cd} from vibrocone tests at two Japanese sites that have undergone several seismic events. Site 1 has historically shown no evidence of liquefaction (e.g., sand boils or surface settlements) following earthquakes. In this case, both static and dynamic tip resistances are relatively similar (except for two thin localized zones about 3 m deep), thus inferring no major liquefaction problems. In contrast, the results of vibrocone tests at site 2, which is known to have liquefied repeatedly, illustrate that the dynamic resistance is considerably reduced at depths between 2 to 5 meters, reflecting the contractive nature of the sand deposit and high likelihood of liquefaction.

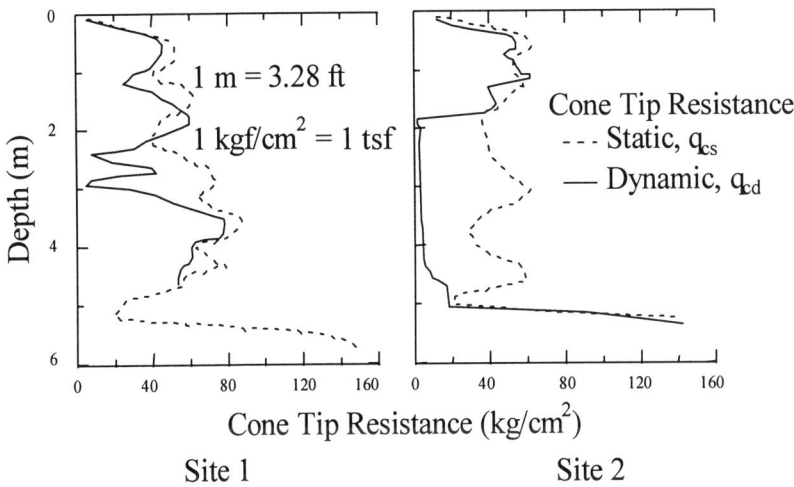

Figure 2. Japanese Vibrocone Results at Two Seismic Sites: (a) Site 1 [not prone to liquefaction]; (b) Site 2 [historicaly-liquefied]. Modified after Sasaki et al., 1984).

A Canadian version of the vibrocone (Moore, 1987) was constructed using an electro-mechanical vertical vibrator applied at the top of the rods. The vibrator consisted of an oscillating pair of eccentric counter-weights to the actuator assembly above hole. The advantage was that the large size vibrator did not need to travel downhole, however, there were likely energy losses with increasing depth. The electric power for the vibrator was coupled with the rig power, causing fluctuations in operating frequency. Limited field tests in a silt showed a reduction in q_c, but a shoulder pore pressure element did not show evidence of excess pore pressures.

Table 1. Contributions to the Development of the Vibrocone

Country	Author(s)	Details	Results
Japan	Sasaki & Koga, 1982 Sasaki et al., 1984 (PWRI)	• Down-hole vibration at 200 Hz • 32 kgf horizontal centrifugal force	Reduction in q_c reflected possible liquefiable zones
Japan	Teparaksa, 1987 (PWRI)	• Down-hole vibration at 200 Hz • 80 kgf horizontal centrifugal force	Compared sister sets of static and dynamic soundings
Canada	Moore, 1987 (UBC)	• Vibration applied at top of rods • Vertical force at 75 Hz frequency	Shoulder pore pressures did not identify liquefiable layers
Italy	Mitchell, 1988 Piccoli, 1993 (ISMES)	• Down-hole horizontal vibration • 200 Hz	Qualitative interpretation of tip resistances
USA	Wise, et al. (1999) Schneider, et al. (1999)	• Downhole vertical-impulses (5 Hz) • Midface porewater pressures	Increased porewater pressures in liquefiable layers

IMPULSE VIBROCONE DESIGN

Under funding from the United States Geological Survey (USGS) and the National Science Foundation (NSF), the development and calibration of a piezovibrocone penetrometer has been initiated in a joint research program at Georgia Tech and Virginia Tech (Wise, et al. 1999; Schneider, et al. 1999). The new vibrocone is an improvement over prior vibrocones for the following reasons:

• Dynamic motion is downhole to prevent energy losses associated with uphole vibrators, as used on the prior UBC version.

• Motion is directed vertically to prevent gapping at the soil-cone interface, since the horizontal centrifugal oscillations of the prior PWRI and ISMES vibrocone penetration tests compromised the quality of the axial CPT measurements.

• Both midface (u_1) and shoulder (u_2) porewater pressures are measured.

• The unit allows for control of dynamic force at lower frequencies (5 to 30 Hz).

Seismic strong motion records from most earthquakes show a spike in their power spectra at frequencies of 1 to 5 Hz, yet previous vibrocones operate at frequencies of 75 to 200 Hz, which are considerably higher. Therefore, from an operational standpoint, a lower frequency vibro-unit was desired having adjustable force excitation.

For the initial design, a downhole pneumatic system was built to offer simplicity in operation and economy in construction. Impulse-type loading was used for the initial trials. Two units were needed, one for field use by Georgia Tech and one for CPT calibration chamber testing at Virginia Tech. For chamber calibration tests, a 15-cm^2 triple-element piezocone was donated by Fugro Geosciences. These tests are ongoing at Virginia Tech (Bonita, et al. 2000). This allows for the simultaneous measurement of cumulative cyclic pore pressures at three locations: midface (u_1), behind the tip (u_2 or u_b), and behind the sleeve (u_3). For initial field trials, a 10-cm^2 Davey-type piezocone with a single midface (u_1) porous element was used.

The fabricated prototype attached to the triple-element piezocone is illustrated in Figure 3. Key resistance parameters are measured, including tip resistance (q_c), sleeve friction (f_s), multiple porewater pressures (u), and frequency content of excitation. Both an HP oscilloscope and spectrum analyzer have been used for the latter.

Two Modes of excitation:

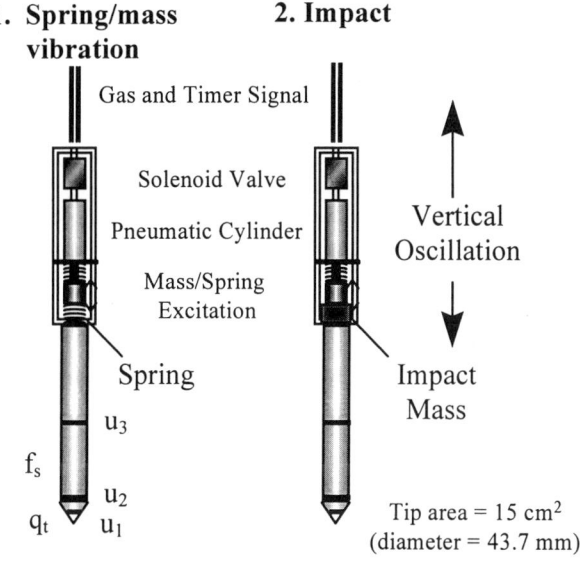

1. Spring/mass vibration   **2. Impact**

Gas and Timer Signal

Solenoid Valve

Pneumatic Cylinder

Mass/Spring Excitation

Spring

u_3

f_s

u_2
q_t u_1

Vertical Oscillation

Impact Mass

Tip area = 15 cm^2
(diameter = 43.7 mm)

Figure 3. Prototype Pneumatic Impulse Generator and Triple Element Piezocone.

The vibratory module consists of a solenoid valve, air cylinder, an impact mass, velocity geophone, a housing assembly, and nitrogen gas. An electronic timer provides electrical pulses to a solenoid valve, which opens and closes at the rate dictated by the timer setting. The solenoid pressurizes the air piston, which in turn drives the excitation. Two modes of excitation are available to increase the versatility of the unit and range of applicability of the results. One mode is driven by a spring-mass oscillation and another uses impulse with an impacting mass. To measure the applied force and frequency, a small OYO Model 14-L9 geophone is installed in the unit.

Without reconfiguring the vibrator, the dynamic force can be adjusted at the control panel by increasing or decreasing the pressure from the tank or compressor. The air cylinder in the prototype has a bore size of 2.7 cm corresponding to a force of 100 kgf at 1.7 MPa pressure. The frequency of excitation can be varied by simply changing the timer setting which is capable of duty cycles from 0.001 seconds to 9999 hours and varied by increments of 0.001 seconds. The solenoid is rated to perform adequately at duty cycles down to 60 per second.

FIELD PERFORMANCE

The impulse vibrocone system has been trial-tested to evaluate its ability in generating excess porewater pressures. The device was advanced in historically liquefiable sands in a series of soundings near Charleston, South Carolina, USA. This seismic region was selected for trial soundings with the piezovibrocone due to the noted abundance of mapped paleoliquefaction features resulting from the Charleston Earthquake of 1886 with Moment Magnitude (M) = 7.5 to 7.7 (Martin & Clough, 1994).

The specific test site locations were based upon the prior documentation of accessible sites where prior SPT, CPT, and grain size information was reported (Martin & Clough, 1990). Gregg In-Situ, Inc. of Aiken, South Carolina provided a 25-tonne cone truck and an electronic 10-cm^2 seismic piezocone with a porous element at the u_2 (shoulder) position. These data provided a baseline reference for comparison with both static and dynamic soundings for the vibrocone penetration test (VCPT).

Two sites have been investigated with the impulse VCPT in the Charleston area (Schneider et al. 1999). One site selected was the Thompson Industrial Services (TIS) in the Atlantic coastal plain region of South Carolina. The area is of minimal relief and low elevation and comprised of sandy sediments deposited between 130,000 to 230,000 years ago. The water table was located about 1.5 m deep. Based on published first-hand accounts and field reconnaissance, the extent of liquefaction due to the Charleston Earthquake had been characterized as extensive at the Ten Mile Hill and as moderate levels at the Eleven Mile Post site (Martin & Clough, 1994)

Static reference soundings were performed with both an electronic ConeTec (type 2) piezocone and the electric Davey (type 1) model. The latter also served as the measurement package upfront for the VCPT sounding during dynamic advancement. The electronic cone provided signal conditioning and amplification of recordings downhole, while the electric penetrometer required filtering and amplification of the signals uphole with the data acquisition system. The tip stresses and sleeve resistances indicate favorable comparisons between the two soundings, as shown by Figure 4. The midface u_1 readings mirrored the shoulder u_2 measurements, yet midface pressures were generally above hydrostatic while shoulder pressures were slightly below hydrostatic conditions. This observation is consistent with static piezocone results reported in clean sands (Robertson, et al. 1986).

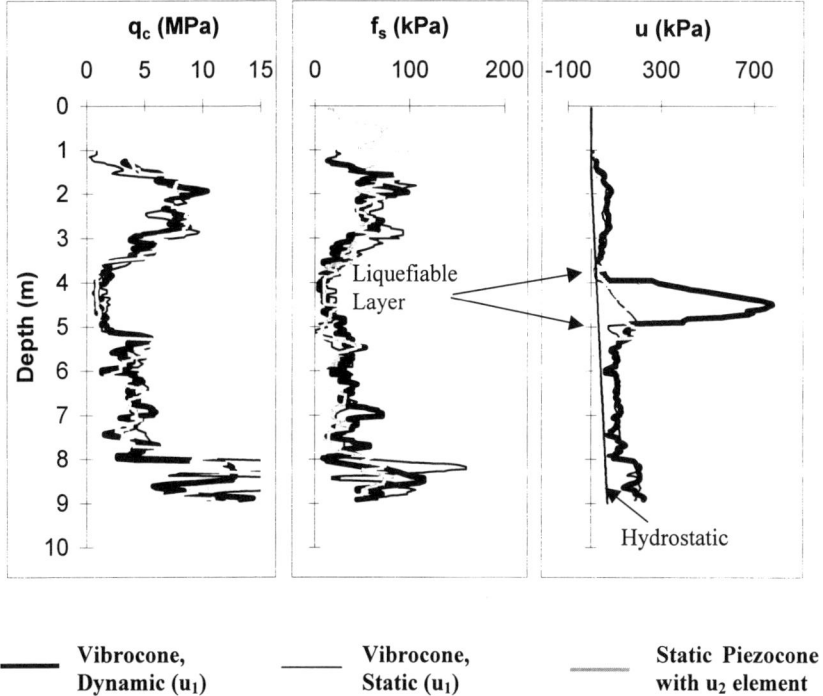

___ Vibrocone, ___ Vibrocone, ___ Static Piezocone
Dynamic (u_1) Static (u_1) with u_2 element

Figure 4. Impulse-Vibrocone Results at the Thompson Industrial Site in Charleston, SC. Readings compared with Reference Type 1 and Type 2 Static Piezocone Soundings

Adjacent to these two static tests, a dynamic sounding was also performed using the vibrocone probe operating at 5 Hz, thus completing a VCPT. At depths of between 3.5 to 5.1 meters, Figure 4 shows a significant spike in the porewater pressure response of the dynamic sounding. This response is interpreted to indicate the zone of liquefiable soil. The recorded tip resistances for all three soundings were relatively low in this region, but the dynamic tip resistance (VCPT) does not noticeably fall below either of the tip resistances from the two static soundings. In the future evaluations, a rational interpretation for the VCPT will consist of effective stress analyses to separate out the mechanisms of static liquefaction, steady-state response, quasi-liquefaction, cyclic mobility, and dilatant behavior of soils (Schneider et al. 1999). The measured q_t and penetration porewater pressure readings, in fact, may reflect totally different portions of the induced effective stress paths.

While the pneumatic vibrocone eliminated some of the problems evident in the prior models, it is not without its own disadvantages. The gas supply line, which must be threaded through the rods together with the penetrometer cable, causes tangling and difficulties in field use. The heavy nitrogen tank is awkward to transport and has to be refilled often. The increased diameter of the trailing vibro-unit required extra pushing force for the cone rig, and created a large diameter hole, which did not provide adequate support for the rods. Buckling in the rods can reduce the amount of force delivered to the tip of the penetrometer, and increases the chances of rods breaking in the hole. Thus the original model was buried in the New Madrid seismic region in December 1998. A second version of the device was constructed that provided dynamic force measurements during advancement using an integrated micro-mechanical sensor (iMEMs) that served as an accelerometer (Hendren, 1999). This model showed that the generated signal and the probe response are not sinusoidal. Also, the air-actuated piston ha a slow response and very limited cycle rate. In order to overcome the shortcomings of the pneumatic-type vibrocone, a new electronically-activated force generator was developed. A piezoelectric stack, termed a piezo-actuator, is ideally suited to solve the pneumatic vibrocone problems.

LINEAR DISPLACEMENT VIBROCONE DESIGN

Piezoelectric actuators are capable of generating great forces at a wide range of frequencies. The unit mounts directly behind the penetrometer so there are no energy losses due to the rods above. The cone is actually de-coupled from the pushing rods, so the rods above, along with the truck act as a reaction force helping input more of the energy into the soil. The device is entirely electric, eliminating the need for a cumbersome pneumatic umbilical chord connected to the surface. The control and excitation of the actuator is handled by spare wires within the existing cone cable. The system response is nearly instantaneous and almost infinitely controllable by a signal generator. There are no more moving parts, making assembly and disassembly greatly simplified. This can increase productivity in the field and reduce downtime. The overall diameter of the device has been decreased dramatically which reduces the potential problems of the previous model. The impact pulse transmitted through the cone has been replaced by an actual vibratory displacement of the cone that will truly strain the soil. This also greatly reduces the chance of damage to the sensitive penetrometer.

A new penetrometer was needed to work with the re-designed vibro-unit. The older Davey cone is powered and monitored via an unaccommodating permanently connected 10-wire shielded cable. The new penetrometer (see Figure 5), from A.P. van den Berg in the Netherlands, utilizes a 16-wire cable terminated with a 16-pin Lemo connector. The penetrometer requires only 12 wires for operation, leaving 4 available wires. The piezo-actuator unit requires only 2 wires, so the other two could be used to monitor a geophone for performing downhole seismic tests or for an accelerometer (or iMEMs) which could be used to measure the amount of dynamic force the cone is imparting into the soil.

This new electro-vibrocone is currently being assembled with a stainless steel module to house the piezo-actuator unit at Georgia Tech. It is hoped that field trials will be conducted at paleoliquefaction sites in the New Madrid/MO and Charleston/SC seismic regions within the next year. In addition to liquefaction evaluations, it is anticipated that the device will aid in the assessment of post-cyclic residual undrained strengths of silts and sands, as well as serve as a tool for evaluating the outcomes and effectiveness of ground modification techniques, such as vibro-compaction, underground blasting, and dynamic compaction.

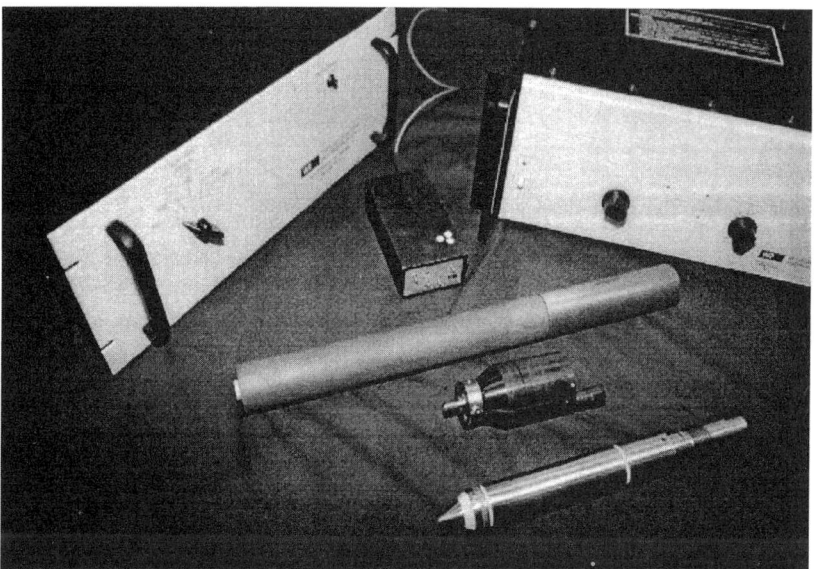

Figure 5. Components of the Electro-Vibrocone Model with Dual-Element Piezocone, Downhole Piezoelectric Shaker, Signal Generator, and Amplifier.

SUMMARY

A new version of a specialized in-situ tool, termed the piezo-vibrocone, has been designed for the direct evaluation of soil liquefaction potential on site-specific projects. The device utilizes a multi-element piezocone coupled with a downhole vibro-unit operating vertically at low frequencies. Trials of the previous pneumatic impulse piezo-vibrocone in historically-liquefied sands of the Charleston seismic region show generation of cyclically-induced porewater pressures. An electronic version of the vibrocone will make significant improvements over the pneumatic version by producing a frequency-controlled sinusoidal displacement without a cumbersome air hose, providing more uniform and repeatable cycles of loading, and multiple porewater pressure data.

ACKNOWLEDGMENTS

The authors thank Dr. Cliff Astill of NSF, Dr. John Unger of USGS, and Dr. Dan Abrams of the Mid-America Earthquake Center for funding on this project. Appreciation is given to R. Yilmaz and R. Klopp at Fugro Geosciences, B. Pemberton at Gregg In-Situ, C. Wise at Black & Veatch, as well as J.K. Mitchell, T.L. Brandon, and J. Bonita at Virginia Tech for their assistance on this project.

REFERENCES

Andrus, R.D. and Stokoe, K.H. (1997), "Liquefaction resistance based on shear wave velocity," *Proceedings*, Workshop on Evaluation of Liquefaction Resistance, Report NCEER-97-0022, Buffalo, NY, 89-128.

Bonita, J.A., Mitchell, J.K, and Brandon, T.L (2000). "The Development and Evaluation of a Vibropiezocone Penetrometer for the Insitu Evaluation of the Liquefaction Potential of Soil Deposits," *Proceedings*, Session on Earthquake Engineering, Geo-Institute Conference/Denver, ASCE, Reston/VA.

Campanella, R.G. and Robertson, P.K., (1988). "Current Status of the Piezocone Test," *Penetration Testing 1998 (ISOPT-1*, Orlando), Vol. 1, Balkema, Rotterdam, 93-116.

Campanella, R.G. (1994). "Field Methods for Dynamic Geotechnical Testing", *Dynamic Geotechnical Testing II*, (STP 1213), ASTM, West Conshohocken, PA, 3-23.

Glaser, S.D., and Chung, R.M., (1995). "Estimation of Liquefaction Potential In-Situ," *Earthquake Spectra*, Vol. 11, No. 3, 431-455.

Hendren, T. (1999). "Second Model of a Pneumatic-Driven Vibrocone Penetrometer", *MS Project*, School of Civil & Enviromental Engineering, Georgia Inst. of Tech., Atlanta.

Kulhawy, F.H., and Mayne, P.W., (1990). "Manual on Estimating Soil Properties for Foundation Design," *Report EL-6800*, Electric Power Research Institute, Palo Alto, 306 pp.

Martin, J.R., and Clough, G.W. (1990). "Implications from a Geotechnical Investigation of Liquefaction Phenomena Associated with Seismic Events in the Charleston, SC Area," *USGS Report*, No. 14-08-001-G-1348, 414 p.

Martin, J.R. and Clough, G.W. (1994). "Seismic Parameters from Liquefaction Evidence". *Jour. of Geotechnical Engrg.* 120 (8), 1345-1361.

Mitchell, J.K. (1988). "New Developments in Penetration Tests & Equipment", *Penetration Testing 1988*, Vol. 1, Balkema, Rotterdam, 245.

Mitchell, J.K., and Brandon, T.L., (1998). "Analysis & Use of CPT in Earthquake and Environ-mental Engrg.," *Geotechnical Site Characterization*, Vol. 1, Balkema, Rotterdam, 69-97.

Moore, D.M. , (1987). "Evaluation of the Cone Penetrometer & Its Effect on Cone Bearing and Pore Pressure," *BS Thesis*, Civil Engineering, Univ. of British Columbia, Vancouver, 73 pp.

Olson, S.M. and Stark, T.D. (1998). "CPT Based Liquefaction Resistance of Sandy Soils", *Geotechnical Earthquake Engineering & Soil Dynamics III*, Vol. One, GSP No. 75, ASCE, Reston/VA, 325-333.

Piccoli, S., (1993). "ISMES Vibrocone," *Personal correspondence* from ISMES, Bergamo, Italy.

Reyna, F. and Chameau, J.L., (1991). "Dilatometer Based Liquefaction Potential of Sites in the Imperial Valley" *Proceedings*, 2nd Intl. Conf. on Recent Advances in Geotechnical Earthquake Engineering and Soil Dynamics, St. Louis, Vol. 1, 385-392.

Robertson, P.K., and Wride, C.E., (1997). "Cyclic Liquefaction Potential and Its Evaluation Based on the SPT and CPT," *Proceedings*, Workshop on Evaluation of Liquefaction Resistance, NCEER-97-0022, Buffalo, NY, 41-87.

Robertson, P.K. and Wride, C.E. (1998). "Evaluating Cyclic Liquefaction Potential Using the Cone Penetration Test", *Canadian Geotechnical Journal*, Vol. 35 (3), 442-459.

Robertson, P.K., Campanella, R.G., Gillespie, D., and Greig, J., (1986). "Use of Piezometer Cone Data," *Use of In Situ Tests in Geotechnical Engineering*, (GSP 6), ASCE, Reston, VA, 1263-1280.

Sasaki, Y., and Koga, Y., (1982). "Vibratory Cone Penetrometer to Assess the Liquefaction Potential of the Ground," *Proceedings*, 14th U.S.-Japan Panel on Wind and Seismic Effect, NBS Special Pub. 651, Washington D.C., 541-555.

Sasaki, Y., Itoh, Y., and Shimazu, T., (1984). "A Study of the Relationship Between the Results of Vibratory Cone Penetration Tests and Earthquake-Induced Settlement," *Proceedings*, 19th Annual Meeting of Japanese Society of Soil Mechanics & Foundation Engrg., Tokyo.

Schneider, J.A., Mayne, P.W., Hendren, T.L., and Wise, C.M. (1999). "Initial Development of an Impulse Piezovibrocone for Liquefaction Evaluation". *Physics & Mechanics of Soil Liquefaction*, A.A. Balkema, Rotterdam, 341-354.

Seed, H.B., and Idriss, I.M., (1971). "Simplified Procedure for Evaluating Soil Liquefaction Potential," *Journal of Geotechnical Engineering*, Vol. 97, No. 9, 1249-1273.

Seed, H.B., Idriss, I.M., Makdisi, F., and Banerjee, N., (1975). "Representation of Irregular Stress Time Histories by Uniform Equivalent Series in Liquefaction Analysis," *Report No. EERC-75-29*, University of California, Berkeley, 40 pp.

Skempton, A.W., (1986). "Standard Penetration Test Procedures & Effects in Sands of Overburden Pressure, Relative Density, Particle Size, Aging, and OCR," *Geotechnique*, Vol. 36 (3), 425-447.

Stark, T.D. and Olson, S.M. (1995). "Liquefaction Resistance Using CPT and Field Case Histories". *Jour. of Geotechnical Engrg.* 121 (12), 856-869.

Suzuki, Y., Tokimatsu, K., Koyamada, K., Tara, Y., and Kubota, Y. (1995). "Field Correlation of Soil Liquefaction Based on CPT Data, *Proceedings, CPT'95*, Vol. 2, Linkoping, Swedish Geotechnical Society, 583-588.

Teparaksa, W., (1987). "Use and Application of Penetration Tests to Assess Liquefaction Potential of Soils," *Ph.D. Dissertation, Department of Civil Engineering, Kyoto University*, Japan, December.

Wise, C.M., Mayne, P.W., and Schneider, J.A. (1999). "Prototype Piezovibrocone for Evaluating Soil Liquefaction Susceptibility", *Geotechnical Earthquake Engineering*, Vol. 2 (Proc., 2nd ICEGE, Lisbon), A.A. Balkema, Rotterdam, 537-542.

MINIATURE PIEZOCONE FOR USE IN CENTRIFUGE TESTING

By Edmundo R. Esquivel, Ph.D.[1] and Claudio H. C. Silva, Ph.D.[2]

ABSTRACT

A miniature piezocone was developed for use in centrifuge environments, including the piezocone design and fabrication. Several penetration tests were performed in centrifuge soil models encompassing normally consolidated and moderately overconsolidated clays, and medium dense to dense sands. For preparation of clay specimens, including preconsolidation, and for performance of piezocone penetration tests (CPTU) in the centrifuge, an experimental apparatus was designed and built. This apparatus allows the performance of tests under different conditions, such as rate of penetration and in-flight test site positioning. This research has shown that piezocone testing in centrifuge models is feasible, producing consistent and reliable data.

INTRODUCTION

The use of centrifuge modeling is a powerful tool for simulating in the laboratory the actual state of stress in the field without the introduction of major perturbation due to boundary effects, which have been the major concern and source of deviations in conventional calibration chamber testing that have been widely used so far. In addition to testing scaled models, it is necessary to verify that the correct soil profile in terms of strength and stiffness has been obtained. Thus, there exists the need for in-flight soil characterization, which can be carried out by performing cone penetration tests or shear vane tests on the soil model being spun in the centrifuge (Ko, 1988).

Piezocones (CPTU), which represents a new generation of in-situ testing devices, are regarded as the most efficient tools for stratigraphic logging of soft soils. Also, they are useful to assess engineering properties of soils, including:
- interpretation of cone resistance in terms of shear strength and deformation characteristics;
- evaluation of *in situ* flow and consolidation characteristics;

---

[1] Geotechnical Researcher, FAPESP, Rua Thomaz Nogueira Gaia 1796, 14020-290 Ribeirao Preto, SP, Brazil. <esquivel@convex.com.br>

[2] Associate Professor, Federal University of Viçosa, 36570-000 Viçosa, MG, Brazil, <silvac@mail.ufv.br>

- evaluation of stress history and overconsolidation ratio of cohesive soils;
- evaluation of soil density and strength characteristics for cohesionless soils;

These factors motivated the development of a miniature piezocone (CPTU), at the University of Colorado, suitable for use in centrifuge testing. Several penetration tests were performed in normally consolidated and moderately overconsolidated clay models, as well as in medium dense to dense sand models. Due to the higher penetration resistance, the performance of penetration tests (CPT) in cohesionless materials required a stronger device. For preparation of clay specimens, including preconsolidation, and for performance of piezocone penetration tests in the centrifuge, an experimental apparatus was designed and built.

CENTRIFUGE MODELING

Scale models in combination with theoretical analyses are frequently used in engineering. Scale modeling is often used when theoretical solutions have to assume major simplifications and approximations, or when numerical solutions are too lengthy or not feasible. This is the case of many geotechnical problems. Geotechnical scale models are also used when building and testing a full scale model is too difficult, dangerous, or expensive.

In geotechnical problems, the soil behavior is governed to a very major extent by the current effective stresses due to acceleration of gravity. Consequently, the stresses at a point in a model should be the same as the stresses at the corresponding point in the prototype. Since the stresses at corresponding points in the model and in the prototype are the same, the soil properties will also be the same (assuming that the stress history in the model is the same as in the prototype) and the behavior of the soil in the model will represent the behavior of the soil in the prototype. The basic principle behind centrifuge testing is to create a stress field in the model that simulates prototype conditions, allowing us to make observations that otherwise could be possible only on full-scale prototypes. The geotechnical engineering problems that can be addressed through centrifuge modeling can be categorized in four groups: modeling of prototypes, investigation of new phenomena, parametric studies and validation of numerical methods (Ko, 1988).

MINIATURE PIEZOCONE

The design of the miniature piezocone CUB1 and penetrometers CUB2 and CUB3 developed at University of Colorado was inspired by the penetrometer Mark II, developed at University of Cambridge (Almeida and Parry, 1984, 1985, 1986). Compared to Mark II penetrometer, the piezocones developed at University of Colorado incorporate three major improvements: (1) development of more sensitive load cells, by using semiconductor strain gages; (2) incorporation of a pore pressure transducer, for monitoring the excess pore pressure developed during the penetration; (3) attachment of the cone to the load cell through threaded connections, so that the cone can be easily replaced. The penetrometers CUB2 and CUB3 have no capability to measure pore pressures.

The piezocone CUB1, schematically shown in Fig. 1, is 305 mm long and 12.7 mm in diameter (D_c). The cone has a projected area of 127 mm^2 and an apex angle of 60°. The piezocone shaft was built from an off-the-shelf aluminum hollow tubing with an external diameter of 12.7 mm and an internal diameter of 6.35mm. All other aluminum parts were machined from solid blocks. Two load cells are mounted at its extremities. The primary load cell, located inside the shaft, measures the point resistance q_c. The pore pressure transducer located inside the primary load cell measures the excess pore pressure developed during the penetration, and its dissipation after the penetration. The secondary load cell, mounted at the base of the shaft, measures the total load applied to the piezocone. The secondary load cell allows an evaluation of the skin friction f_s along the piezocone shaft.

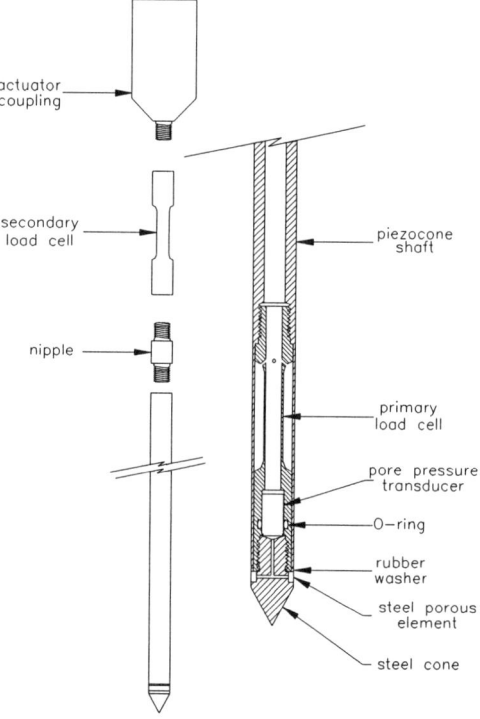

Fig. 1. Schematics of Piezocone CUB1

Since one of the major applications of piezocone testing is assessing properties of soft and very soft clays, a main concern is the sensitivity of the load cells. Considering this fact, the CUB1 primary load cell was designed to measure the relatively low penetration resistance of soft clays. It has a resolution of 0.0005 kN for the range 0-0.5 kN. The high sensitivity is achieved by using semiconductor strain gages. The CUB2 and CUB3 penetrometers, intended to perform penetration tests in cohesionless soils, are provided with off-the-shelf higher capacity load cells.

The piezocone pore pressure transducer is provided with a silicon diaphragm, which ensures very fast response time. The pressure transducer is placed inside the primary load cell, as shown in Figure 1. The filter is located at the cone shoulder. Whereas the piezocone shaft and the load cells are made of 7075-T6 aluminum, the cone is made of stainless steel, and the filter is made of sintered stainless steel. The filter element is obtained from cutting slices from a 10-micron porosity filter cup, with an external diameter of 12.7 mm and an internal diameter of 9.5 mm.

THE PENETRATION TESTING APPARATUS

The basic framework of the penetration testing apparatus consists of a steel base, two welded steel support frames, one main frame and one auxiliary frame, and a rotary stand, which supports the model container. The steel frames are used to support the pneumatic loading system during the consolidation process, to help placing the whole apparatus on the centrifuge platform, and to support the piezocone driving device (Fig. 2).

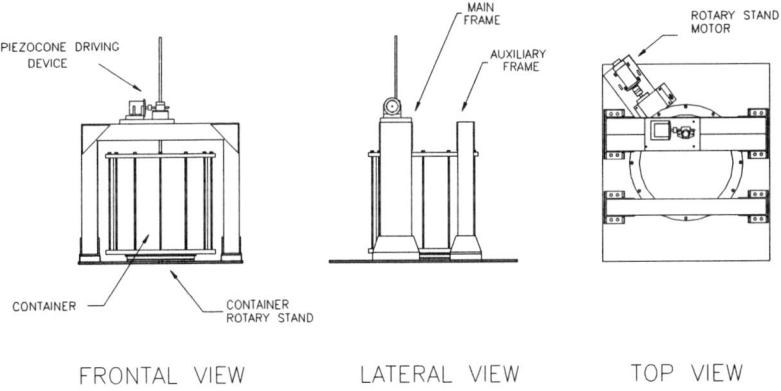

Fig. 2. Penetration Testing Apparatus

The purpose of the container rotary stand is to allow different penetration tests at different sites in the same tub of soil, without stopping the centrifuge to change the position. The rotary stand consists of a steel disk pivoted in the center and supported by hard metal ball bearings. The rotary stand is driven by a servomotor, through a gear reducer (1:60), a chain and sprockets.

The piezocone driving device consists of a linear mechanical actuator, which is driven by a servomotor. The position and the motion of the piezocone are controlled from a computer located in the centrifuge control room.

MOTION CONTROL SYSTEM

A modular motion control system remotely controls the operation of the piezocone driving device, the container rotating stand and the vane apparatus. The basic system is made up of four components: (1) the host computer, from which the commands to the drive system originate; (2) the indexer, which receives commands from the host computer and generates the pulse stream required to control the motor and drive systems; (3) the servo amplifier, which provides the necessary power amplification to maintain the operation of the motors; and (4) DC servo motors and optical encoders with two channels in quadrature. The optical encoders connected to the motors, provide feedback signals to control position, speed, acceleration, direction and range of the movement. The menu-driven program MOTOP is used to operate the motion control system. Through this software, the user selects the option to be executed and is prompted for inputs that define the direction, range and speed of motion.

PENETRATION TESTS IN CLAY

A series of piezocone penetration tests were performed in soil specimens prepared by consolidating a slurry made of Speswhite fine china kaolin. For reference, the physical index properties of this material are listed in Table 1. The specimens were prepared to represent typical soil profiles. Each specimen was prepared in such a way as to obtain a specified overconsolidation ratio at the gravity level under which the penetration tests were performed.

Table 1. Index Properties of Speswhite China Kaolin.

Percent finer than 2 microns (by weight)	75%
Specific gravity, G_s	2.66
Liquid limit, W_l	53.1
Plastic limit, W_p	31.9

The penetration tests were performed at seven different sites along a circumference with a radius of 180 mm and 45° apart. In this way, the distance between two adjacent penetration sites was 130 mm, which corresponds to 10.8 piezocone diameters. The shortest distance from any penetration site to the container wall was 106 mm.

The obtained resistance profiles showed good repeatability, meaning that any eventual boundary effect or interference from adjacent tests can be neglected while interpreting penetration test data.

Figures 3, 4, 5 and 6 show the plots of the corrected penetration resistance (q_T) versus depth for four different soil samples. Each curve represents the penetration resistance profile for one penetration site. Also, are shown the plots of the pore pressure generated at the cone shoulder (u_T) during the penetration. The corrected penetration resistance was obtained with the aid of the following expression (Lunne et al., 1997):

$$q_T = q_c + u_T \ (1\text{-}a)$$

where q_c is the total penetration resistance, and a is the net area ratio ($a = 0.56$). It can be noticed that for each specimen, the curves representing each tip resistance profile are close to each other, meaning that the tests show a good degree of repeatability.

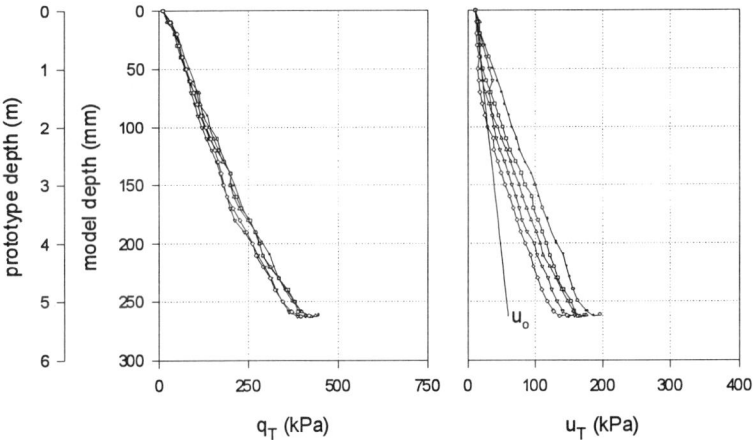

Fig. 3. CPTU Performed in Overconsolidated Clay (6<OCR<14, g-level=20).

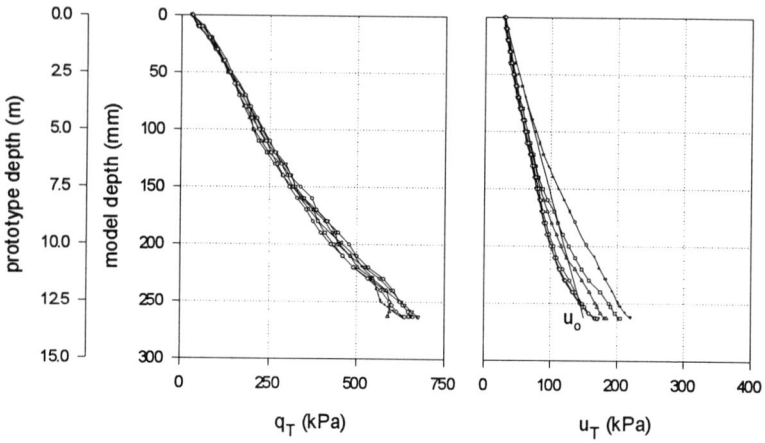

Fig. 4. CPTU Performed in Overconsolidated Clay (4<OCR<10, g-level=50).

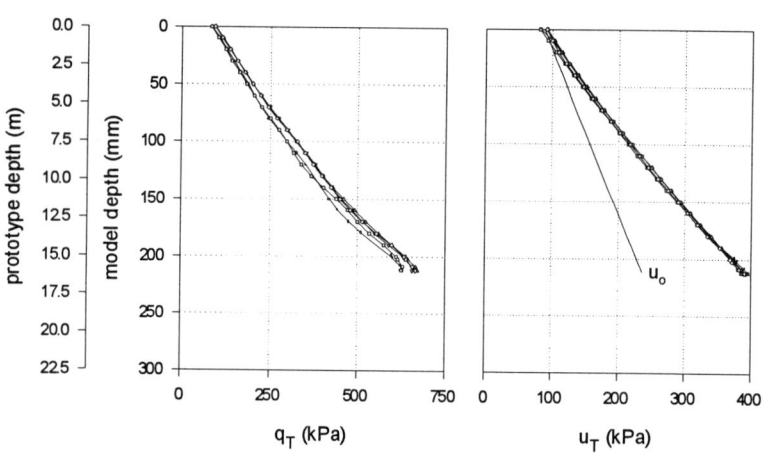

Fig. 5. CPTU Performed in Normally Consolidated Clay (g-level=75).

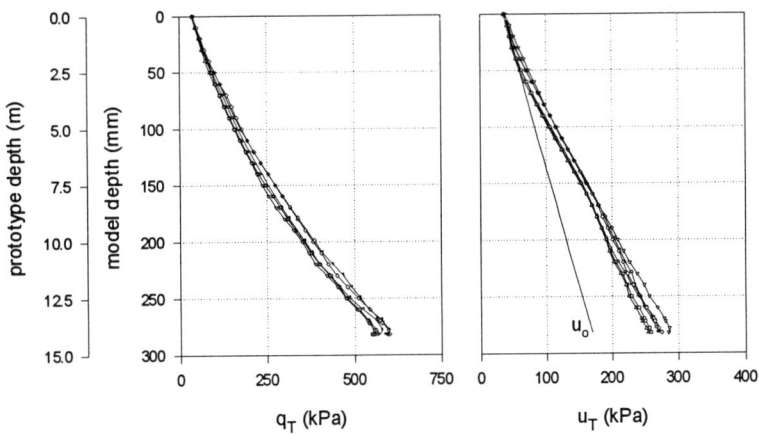

Fig. 6. CPTU Performed in Normally Consolidated Clay (g-level=50).

PENETRATION TESTS IN SAND

Ideally, the cone penetration testing in cohesionless soil models should be performed under no influence of *boundary effects* and *scale effects*. The boundary effects can be caused by the following factors: development of shear stresses in the contact between the sand specimen and the container wall; arching effects caused by the friction that is developed between the specimen and the container wall and influence of the rigid bottom plate on test results. The main scale effects are related to the ratio between the average grain size diameter (D_{50}) and the cone diameter (D_c).

In order to investigate the performance of cone penetration testing in centrifuge models, a series of penetration tests were performed in sand specimens submitted to different test conditions. The main purpose of these tests was to investigate boundary effects and scale effects, and how they interfere on CPT test results. The tests were performed with samples prepared with Nevada #100 Sand. For reference, its physical properties are listed in Table 2.

Each sample was prepared by the pluviation technique to a specified relative density. After checking its volume and weight, the sample was placed in the rotary stand for being tested in the centrifuge. Figure 6 depicts a top view of the container showing the positions where the penetrations tests were performed. By means of the rotary mechanism, any position along the outer circle (Position A) and along the inner circle (Position B) could be reached in flight.

Table 2. Nevada #100 Sand Properties

Specific gravity, G_s		2.67	
Max. dry density, $\rho_{d,max}$		17.33 kN/m³	
Min. dry density, $\rho_{d,min}$		13.87 kN/m³	
Max. void ratio, e_{max}		0.89	
Min. void ratio, e_{min}		0.51	
Avg. grain size		0.15 mm	
Relative density, D_r	58 %	72 %	85 %
Dry unit weight, γ_d	15.68 kN/m³	16.19 kN/m³	16.70 kN/m³
Peak friction angle, ϕ_{peak}	37°	41°	45°

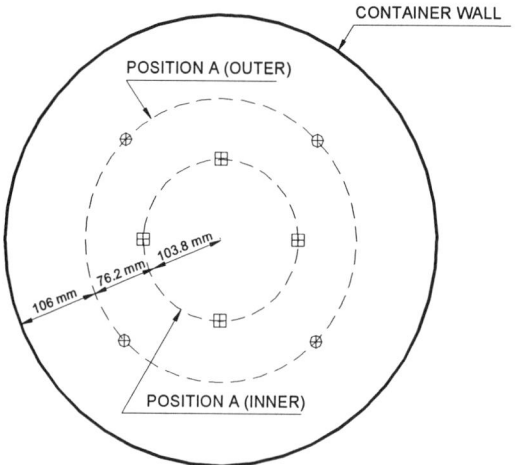

Fig. 7. Penetration Test Site Position

For cohesionless soils, one of the most important aspects that may interfere with CPT results is related to *scale effects*. The importance of these effects is mainly related to the ratio between the cone diameter (D_c) and the average grain size diameter (D_{50}) of the specimen sand. Performing penetration tests in calibration chambers, de Lima and Tumay (1992) suggested correction factors to adjust the measured penetration resistance (q_c) achieved with cones having a diameter different from 35.7 mm.

Another source of concern are the so-called *boundary effects*, which are the effects of the imposed boundary conditions on test results. When performing penetration tests in conventional calibration chambers, the rigid testing container wall and rigid plates used to apply the vertical overburden pressure may cause perturbations, which have to be accounted for in the results interpretation. As observed by Parkin (1988), potential boundary effects are governed by the calibration chamber diameter (D_{cc}), the specimen density, the penetrating cone diameter and the distance between the penetration test site and the container wall (L_w).

In the present work, penetration tests were performed to investigate some specific experimental conditions affecting the cone penetration testing. These tests were grouped in two series, which are briefly described as follows.

The first test series was intended to investigate *scale effects*. Penetration tests were performed in the same specimen with cones having different diameters. Each specimen having a specified density was tested under different g-levels. Special attention was given to investigate the variation of the cone penetration resistance with the cone diameter. Figure 8 shows typical results obtained in this test series. It can be seen that the penetration resistance profiles, obtained with penetrometers with different cone diameters, are very close to each other.

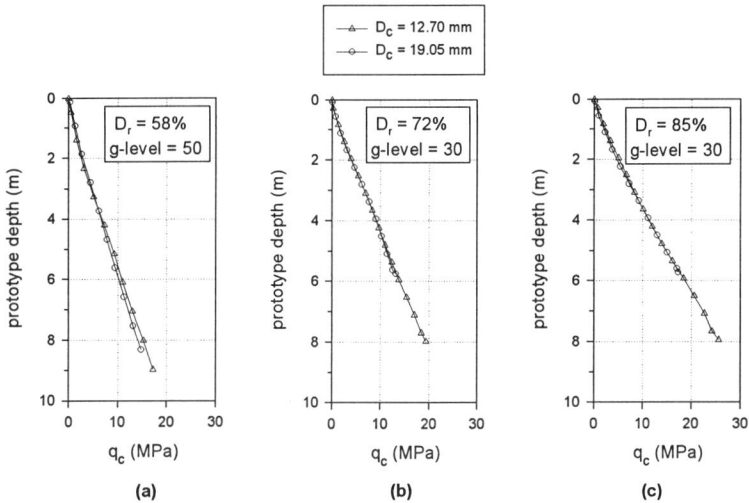

Fig. 8. Penetration Tests Results Showing Negligible *Scale Effects*

The second test series was intended to investigate *boundary effects*. Initially, it was investigated the influence of the test site position from the container wall on test results. Penetration tests were performed in the same specimen, with the same penetrometer, at sites with different distances from the container wall. Figures 9a and 9b show comparisons when the tests were performed at sites located along the inner and outer paths, as shown in Figure 7. The next step in this test series was to search eventual arching effects caused by the friction between the sample and the container wall. In order to minimize the development of shear stresses in the contact between the sand specimen and the container wall, the specimen was involved by a latex membrane soaked with oil. Also, it was studied the influence of the rigid container bottom plate on the penetration test results. This influence was searched by installing a semi-circular 20 mm thick foam rubber plate on the container bottom. Since arching effects are more prone to be noticed at higher densities, the specimen was molded at a relative density equal to 85%. Figure 9c show a comparison between penetration resistance profiles, corresponding to specimens with membrane and without membrane, respectively.

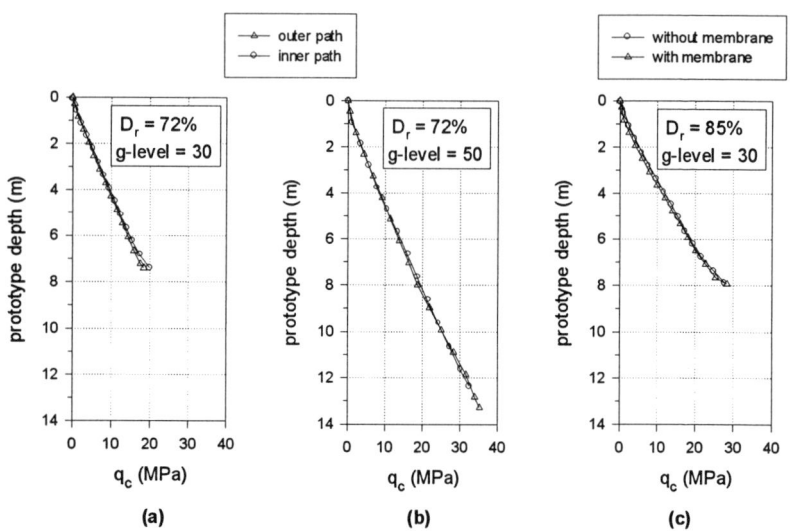

Fig. 9. Penetration Tests Results Showing Negligible *Boundary Effects*

CONCLUSIONS

This research has shown that CPTU and CPT testing in centrifuge environments is feasible, producing consistent and reliable data. With the aid of a centrifuge, it is possible to test models under conditions that approximate those in the field in terms of in-situ stress profile, including the vertical effective stress gradient. Cone penetration testing results in the centrifuge environment are practically not affected by boundary conditions. Piezocone tests performed on centrifuge soil models may produce a large data base leading to the study of correlations such as clay undrained shear strength versus penetration resistance or sand peak friction angle versus cone penetration resistance. Penetration tests in sand centrifuge models have shown no evidence of *scale effects* for the ratio D_c/D_{50} ranging from 84 to 127. Also, these tests have shown no noticeable evidence of *boundary effects* for the ratio D_{co}/D_c ranging from 30 to 45 and the ratio L_w/D_c ranging from 5.6 to 14.3.

ACKNOWLEDGMENTS

The authors are extremely grateful to Prof. Hon-Yim Ko for his constant support, guidance and friendship. The financial support provided by Coordenação de Aperfeiçoamento de Pessoal de Nível Superior (CAPES), Conselho Nacional de Pesquisa (CNPq) and the National Sciences Foundation, gratefully acknowledged, was essential to the success of this research. In addition, the first author is grateful to the Fundação de Amparo à Pesquisa do Estado de São Paulo (FAPESP) for allowing the continuation of this research.

REFERENCES

Almeida, M.S.S., Parry, R.H.G. (1985), Small cone penetrometer tests and piezocone tests in laboratory consolidated clays, *Geotechnical Testing Journal*, GTJODJ, Vol. 8, No. 1, pp. 14-24.

Almeida, M.S.S., Parry, R.H.G. (1984), Penetrometer apparatus for use in the centrifuge during flight, *Symposium on the Application of Centrifuge Modelling to Geotechnical Design*, Manchester, UK, pp. 47-65.

Almeida, M.S.S., Parry, R.H.G. (1988), Miniature vane and cone penetration tests during centrifuge flight, *Vane Shear Strength Testing in Soils: Field and Laboratory Studies*, ASTM STP 1014, A.F. Richards, Ed., American Society for Testing and Materials, Philadelphia, pp. 209-219.

de Lima, D.C. and Tumay, M. T. (1992), Scale effects in cone penetration tests, *ASCE Geotechnical Engineering Congress*, ASCE Special Publication N° 27, GT DIV./ASCE, Boulder, CO, pp.38-51.

Esquivel, E.R. (1995) *Piezocone Testing: Centrifuge Modeling and Interpretation*. Ph.D. Thesis, University of Colorado at Boulder, Boulder, Colorado, 260p.

Ko, H.Y. (1988). Summary of the state-of-the-art in centrifuge model testing. *Centrifuges in Soil Mechanics*, W.H. Craig, R.G. James and A.N. Schofield (eds), Balkema Publishers, Rotterdam, pp. 11-18.

Lunne, T., Robertson, P.K., and Powell, J.J.M. (1997) *Cone Penetration Testing in Geotechnical Practice*, Blackie Academic and Professional, London, 312 p.

Parkin, A. K. (1988) The calibration of cone penetrometer, *First International Symposium on Penetration Testing 1988*, Vol. 1, Balkema Publishers Rotterdam, pp. 221-244.

Renzi, R., Corté, J.F., Bagge, B., Gui, M., Lane, J. (1994), Cone penetration tests in the centrifuge: experience of five laboratories, *Centrifuge 94*, Balkema Publishers, Rotterdam, pp. 77-84

Silva, C.H.C., (1998) *Centrifuge Modeling of Cone Penetration Testing in Cohesionless Soils*, Ph.D. Thesis, University of Colorado at Boulder, Boulder, Colorado, 226p.

INCLUSION OF THE PERFORMANCE MODEL
TO DIRECT AND CONTROL SITE CHARACTERIZATION

C.H. Dowding[1], H.W. Reeves[2], A.J. Graettinger[3], J. Lee[4]

ABSTRACT

A powerful method to quantitatively control and direct exploration that combines the uncertainty (or variance) of three dimensional (3D) site characteristics with sensitivity of the performance model through a Taylor series expansion is demonstrated. Fundamentally, the approach is made possible by directly differentiating the performance model to obtain its 3D sensitivity to changes in subsurface properties. A one-time only direct differentiation of a generic code for performance analysis allows sensitivity to be determined with a single model run for any subsurface geometry. Thus direct differentiation avoids 1000's of model runs necessary with parameter perturbation or Monte Carlo simulation.

The integrated exploration approach will be demonstrated with three examples, each with very different performance objectives (settlement, groundwater flow, and contaminant transport). The examples will demonstrate the process of producing calculated performance variance from 3D model sensitivity and uncertainty associated with site layer geometry and properties. Exploration is directed to locations of maximum variance in project performance. Sampling at these points produces the maximum reduction in performance uncertainty. As directed exploration proceeds, sufficiency of characterization effort is quantitatively determined through a reliability index. This index combines calculated performance, variance in calculated performance, and required performance.

INTRODUCTION

Site characterization is often incorrectly thought to have as its objective the definition of subsurface layer geometry and material properties. Such an objective is too

---

[1] Professor, Dept. of Civil Eng., Northwestern University, Evanston, IL, 60208

[2] Assistant Professor, Dept. of Civil Eng., Northwestern Univ., Evanston, IL, 60208

[3] Assistant Professor, Dept. of Civil and Env. Eng., Univ. of Alabama, Tuscaloosa, AL, 35487

[4] Research Assistant, Dept. of Civil Eng., Northwestern Univ., Evanston, IL, 60208

broadly conceived, as characterization is not conducted as an end onto itself but rather to provide information in order to make a decision. Without a strict inclusion of the goal engineering, regulatory, etc., characterization is normally controlled by answering the following questions. Was the best available technology employed? Were typical resources allocated to exploration? Were all historical resources included in the characterization? Inclusion of the performance model in the characterization process provides the only rational basis to quantitatively control and direct site characterization activities.

Optimizing the location of the small number of borings employed during site characterization is becoming increasingly important because the need for more precise characterization conflicts with the limited availability of exploration resources. Typically less than one millionth of a site volume is sampled during exploration. Therefore, engineering and geologic judgment are called upon to leverage this information to direct exploration and prepare models for analysis. The need for a more rigorous, less subjective, probabilistic approach to site exploration for geotechnical engineering projects has been discussed for decades.

This paper describes an efficient quantitative method of selecting boring locations called quantitatively directed exploration (QDE). As shown in column 2 of the calculation flow chart in Fig. 1, this method combines 1) uncertainty of site characterization data, and 2) sensitivity of the project performance model, to locate the next boring at the position of greatest importance. Importance, which is the estimated

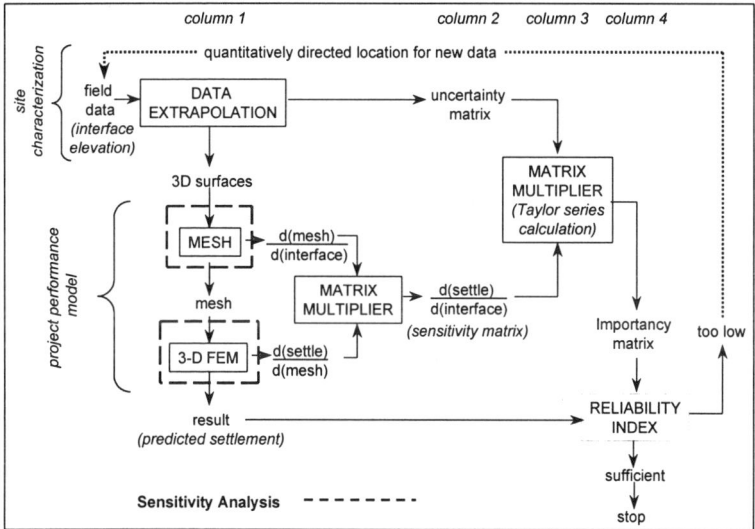

Fig. 1. Flow chart of subprograms (boxed and capitalized) and data matrices (unboxed and lower case) required to produce the importance matrix employed in directed exploration (after Graettinger and Dowding, 1999).

variance in calculated performance (Tomasko et al., 1987, Ditmars et al., 1988), is the product of the model sensitivity and input data uncertainty. Through the 3D importance matrix, QDE combines engineering analysis and geologic characterization into a single, three-dimensionally distributed variance parameter for directing exploration.

Importancy or variance in calculated performance is needed because neither the performance sensitivity (greatest near the largest loads) nor the characterization uncertainty (greatest at a distance from boring locations) alone indicates the optimal location for the next boring. In the QDE method described herein, information gathered from the optimal boring location will yield the greatest reduction in calculated performance variance. The fundamental mathematics of QDE, based on a first-order Taylor series expansion, are described elsewhere (Graettinger and Dowding, 1999).

The reliability index, shown in column 4 in Fig. 1, is employed to determine when exploration is adequate. It is produced from a combination of: a) the expected value of calculated performance (predicted settlement -bottom of column 1), b) the variance in calculated performance (importance matrix, column 4), and c) a predetermined performance goal. Comparison of the performance goal and the probability distribution of calculated performance allows for the calculation of project reliability. If the reliability index is sufficiently high, given the current subsurface uncertainty, then exploration stops. If it is too low, more exploration is required. Details of the reliability index computation for settlement examples is presented in Graettinger and Dowding (2000).

Since the QDE approach is based on a 3D finite element model employed to calculate project performance, it can be employed for a large variety of subsurface performance analyses. This paper describes three of these applications: settlement, groundwater flow, and contaminant transport. The examples are performed on layered 3D domains. The settlement example involved true 3D computation of displacement. A horizontal flow equation is solved reducing the dimensionality of the governing equation for the flow example. Contaminant transport computations are performed on the same domain used in the flow example, but include the extra dimensionality of time. Project performance for each of these cases is contingent upon changes in the geologic interface location and layer properties, which are dealt with independently in the QDE approach and in this paper. QDE can be extended to other subsurface analyses involving 3D FEM solutions, such as soil/rock-structure interaction.

DIRECTED EXPLORATION FOR SETTLEMENT ANALYSIS

Operation of QDE to locate the next boring is demonstrated with actual field data at a site where limiting surface settlement is the design objective. The site is located on thick alluvial deposits consisting of silty clay and dense sands overlying stiff to very stiff clays (Finno, 1996). Five existing soil borings (SB-1 to SB-5) drilled to a depth of 15.25 meters define the subsurface. Design loads are 23 kPa for the mat and 48 kPa for the perimeter strip footings, and these loads are applied as nodal loads in the finite element model.

Interface elevations of four geologic layers are shown by the cross section in Fig. 2a. Soil properties of Young's Modulus (E) and Poisson's Ratio (ν) for each layer were determined from soil samples and used as input for the 3D FEM. Table 1 shows soil parameters for the four layers at this site. While it is possible to incorporate uncertainty of layer properties by adding another set of terms to the Taylor series expansion (Baecher and Ingra, 1981), layer properties are held constant in this example to focus attention on the spatial uncertainty of the interface locations.

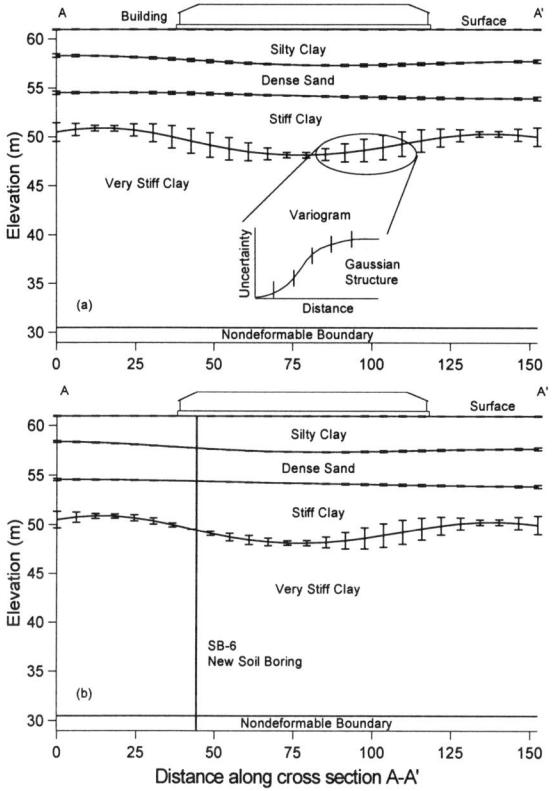

Fig. 2. (a) Cross section A-A' including building location, geologic layer interfaces, and interface uncertainty error bars. Inset shows a variogram indicating how interface uncertainty is reduced near boring locations. (b) Cross section A-A' after data from QDE boring SB-6 is included.

A spatial correlation structure based upon interface elevations at the case study borings was employed to extrapolate interfaces. The correlation structure was determined by fitting typical variogram shapes (Isaaks and Srivastava, 1989) to the site specific data of interface elevation differences and borehole separation distance.

Different interfaces were considered to be statistically independent. Both linear and Gaussian variogram shapes were employed in this example. The Gaussian shape is shown in the inset in Fig. 2a. Vertical bars on the variogram inset indicate the horizontal dimensions of the finite elements. This spatial correlation for soils is similar to that reported by others (Lacasse and Nadim, 1996).

TABLE 1. Soil layer properties

Layer	Young's Modulus[1], E (kPa)	Poisson's Ratio, ν
Silty clay	6,420	0.3
Dense sand	26,350	0.3
Stiff clay	18,400	0.3
Very stiff clay	22,810	0.3

[1] from constrained modulus and assumed Poisson's ratio

DIRECTING AND TERMINATION OF EXPLORATION EFFORTS

Total importancy, which is the variance of calculated surface settlement, is employed to determine the location for the next boring. Importancy is produced by first multiplying the sensitivity of surface settlement to changes in interface elevation by the associated elevation uncertainty and then summing these values for all four interfaces to produce a total importancy or settlement variance at each x, y location.

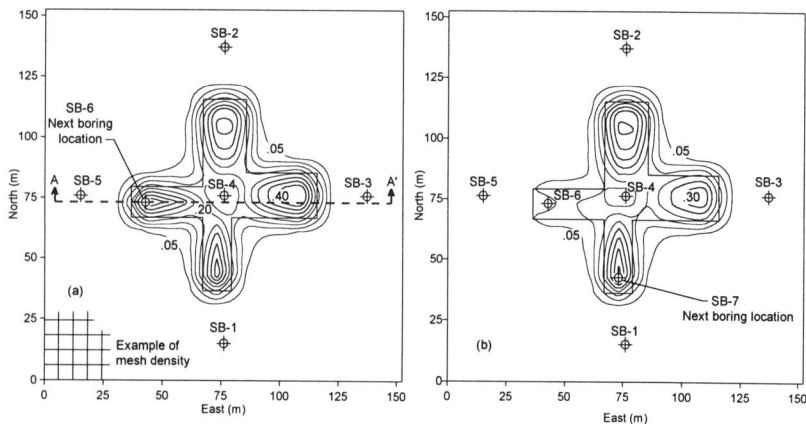

Fig. 3. (a) Contour map of the square root of total importancy (mm) indicating the next boring SB-6, the building, and cross section A-A'. (b) New importancy distribution after inclusion of data from SB-6, along with the location of next boring SB-7

For this example, the location of greatest importance or variance is employed to identify the optimal location for the next soil boring, labeled SB-6 in Fig. 3a. Fig. 3a is a contour plot of the total importance produced from data from the initial five borings. The asymmetry of importance results from the asymmetry of both the building footprint (Fig. 3) and the interface elevations and variance (Fig. 2). Assumed interface elevation data were added to the original site information as if SB-6 were drilled and the QDE analysis was rerun to produce a new total site importance, shown in Fig. 3b. The same approach used to locate SB-6 was then employed to locate the next sampling point, SB-7, which is shown in Fig. 3b.

2D STEADY STATE GROUNDWATER FLOW MODEL

QDE is applied to two dimensional groundwater flow model. The plan view of the aquifer system is shown in Fig. 4. A regional piezometric head gradient is set across the site as shown on the figure, and the transmissivity varies linearly from 50m²/day along the left-hand side to 100m²/day along the right-hand side. The transmissivity at each node spatially varies as the aquifer thickness changes, and the conductivity of 8m/day is assumed to be constant in the aquifer. Only transmissivities are used as the uncertain input parameters, which are related to piezometric head elevations to obtain the model sensitivity. At the 4 sampling locations(SB-1 to SB-4) on the figure, observations for layer thickness (transmissivity/conductivity) are obtained to be used in calculation of variation in the geological model. A proposed extraction well with a pumping rate of 500 m³/day is located as shown on the figure to generate a capture zone for a contaminant plume.

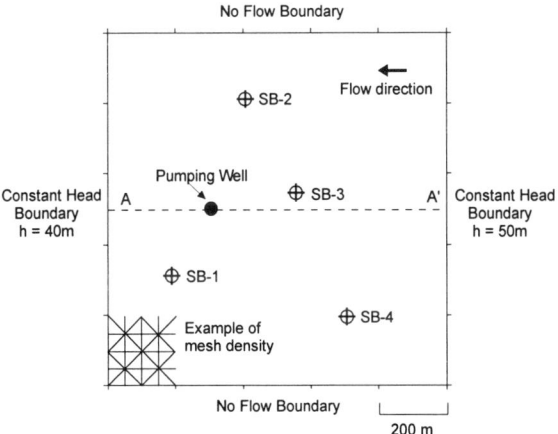

Fig. 4. Plan view of ground water flow geometry including borings, pumping well and pressure head boundary conditions

Fig. 5 (a) and (b) show the expected aquifer thickness, which is obtained from the expected transmissivity input with constant conductivity assumption, and its estimated standard deviation. Since no nugget effect is assumed in the autocorrelation function, the standard deviation tends to zero at the measurement points and to its maximum value away from these points.

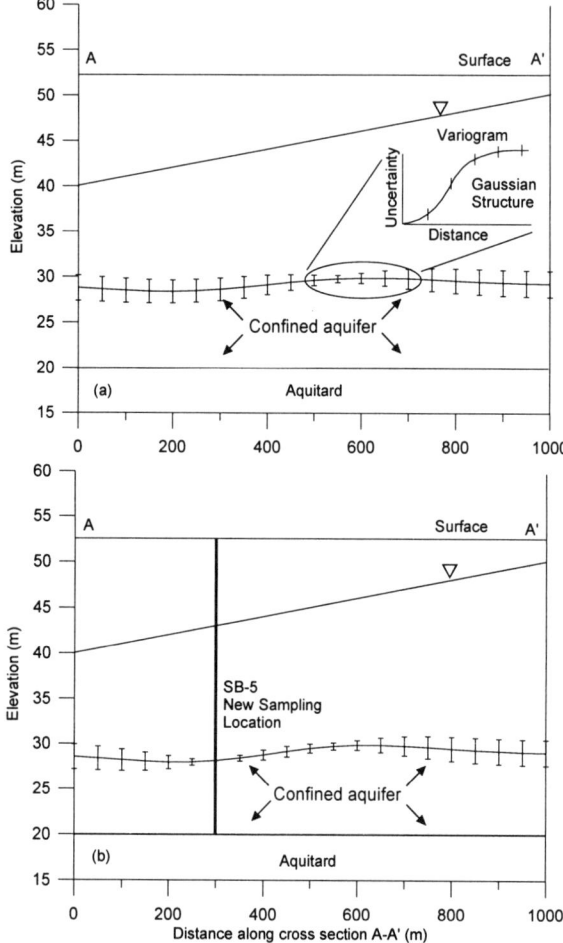

Fig. 5. (a) Cross section A-A' including confined aquifer and its uncertainty error bars for the thickness. (b) Cross section A-A' showing the location of the new sampling at extraction well and its effect on the uncertainty.

To account for system design in the site exploration process, the behavior of the design model must be considered. To get spatially-correlated variance or importancy of the head, shown in Fig. 6, the matrix of the uncertainty of the input parameter is multiplied by the sensitivity of the performance model. The location with the maximum value (the pumping well) indicates the most important location to collect data, which in this case is layer thickness. As with the settlement example, obtaining layer thickness at this maximum location reduces local and overall uncertainty. After addition of layer thickness at SB-5, the procedure is repeated to obtain the next most important location. For this example, Fig. 6(b) shows the location of the next most important sampling point, labeled SB-6. This location is where the uncertainty in estimated transmissivity is large and the influence of the pumping well is still present.

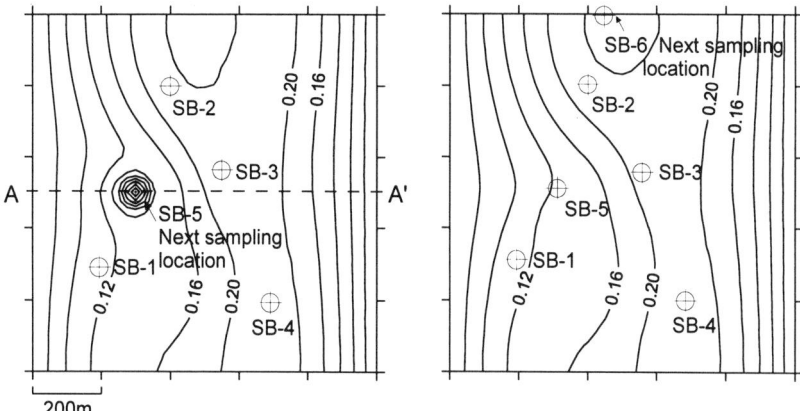

Fig. 6. (a) Contour map of the square root of importancy of head (m) indicating the next sampling location. (b) New importancy distribution after inclusion of additional data at SB-5.

TRANSIENT CONTAMINANT TRANSPORT

The previous two examples, settlement and groundwater flow, relied upon steady-state design models. The QDE procedure, however, is also applicable to transient problems. Consider the case of contaminant transport in the uncertain flow field modeled in the previous section. An initial pollutant plume, which may have uncertainty in regards to maximum concentration and extent, is shown in Fig. 7(a). Snapshots of the plume as it migrates towards the pumping well for 500 and 1000 days of travel are shown in Fig. 7 (Reeves et al., 1999). If we consider the sensitivity of our concentration estimates with regards to the uncertain transmissivity, a time-dependent importancy may be computed. To illustrate the importancy changes, Fig. 8 shows a contour of the diagonal of the importancy matrix for 500 and 1000 simulation times. The matrix is diagonally dominate, therefore these contours show the behavior of the uncertainty. Note that the position of the maximum importancy changes as the solute migrates. Also note that the location of maximum importancy tends to occur at locations of maximum concentration change, not at the peak concentration of the plume. By considering the time-varying importance, exploration may be directed and the time-dependent reliability of remediation schemes may be estimated.

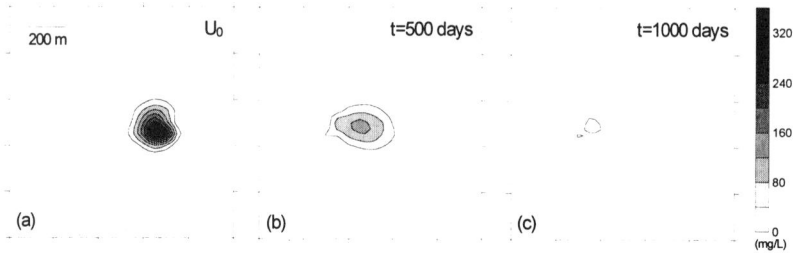

Fig. 7. Snapshots of the plume migration: (a) Initial pollutant plume, (b) Plume traveled for 500 days, (c) Plume traveled for 1000 days.

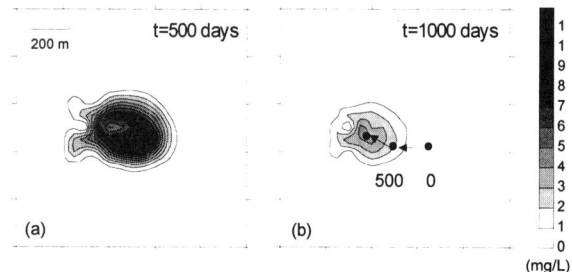

Fig. 8. Contour map of the square root of diagonals of importance matrix. (a) Importance for 500 days simulation time (b) Importance for 1000 days simulation time

IMPORTANCE GOVERNS PERFORMANCE-DRIVEN DECISIONS

As shown by these three examples, neither uncertainty in interface location nor model sensitivity alone can adequately direct or evaluate exploration. In the QDE approach these two considerations are combined through a first-order Taylor series expansion to produce an importance distribution, which mathematically is the variance in computed model results. The importance at any node is influenced by all interface elevations at the site because each interface elevation affects system response in these examples. Unlike sensitivity, which changes only when the best estimate of the interface elevation changes after the addition of a new boring, importance responds significantly to the new information because of the reduction of uncertainty around the boring.

A first-order Taylor series expansion for correlated input data is shown by Equation 1 (Townley and Wilson, 1985),

$$Cov\left[Y_k, Y_l\right] \approx \sum_{i=1}^{n} \sum_{j=1}^{n} \left(\frac{\partial \ f_k}{\partial \ \bar{x}_i}\right)\left(\frac{\partial \ f_l}{\partial \ \bar{x}_j}\right) Cov\left[\bar{x}_i, \bar{x}_j\right] \tag{1}$$

where $Cov[Y_k, Y_l]$ is the covariance or importance of the performance parameter between node k and node l. For the settlement example there are 676 surface nodes. $(\partial f_k/\partial x_i)$ is the sensitivity of surface node k to the elevation of interface node i; since there are 676 surface nodes and 676 elevation nodes on each interface, there are over $(676 \times 676) = 4.50 \times 10^5$ surface sensitivities for each interface. $Cov[x_i, x_j]$ is the covariance between elevations of interface nodes i and j. There are 676^2 pairs of nodes on each interface or over 4.50×10^5 covariance terms.

The Taylor series expansion method for calculating importance has limitations. First, the derivatives are calculated at the expected value of interface elevation; therefore, if there is a large uncertainty associated with the interface elevation the derivative may not be equal to the derivative of the true interface elevation (Harr, 1996). Second, in these examples, the interface elevations on one interface were assumed to be statistically independent of the other interfaces in the subsurface. Finally, the Taylor series expansion method requires knowledge of the uncertainty associated with each input parameter. In these examples, a Bayesian framework is used to interpolate the interface elevations and uncertainties at the sites.

DISCUSSION

In each of these three examples, the estimated covariance of the design parameter: settlement, piezometric head, or concentration, is referred to as the importance matrix. Analysis of this matrix is used to direct exploration. Each case considers one spatially-correlated uncertain input to the design model, but the method can be readily extended to consider other input parameters and their correlation structure. The method outlined in this paper is general and relates site exploration to both design model sensitivity and input parameter uncertainty. In addition to directing exploration, the analysis outlined in this paper may be used to quantitativley judge when site exploration is sufficient. Column 4 in Fig. 1 presents the reliability index, which combines model performance, input data uncertainty, and required performance, shown by Equation 2,

$$\beta = \frac{u_{model} - u_{allowed}}{\sigma_{estimated}} \tag{2}$$

where β is the reliability index, u_{model} is the calculated performance, $u_{allowed}$ is the performance goal, and $\sigma_{estimated}$ it the performance uncertainty. This index can be used to judge whether more field data are required for a given design. In other words, when is enough, enough? Fig. 9 shows typical ranges for reliability of several types engineering projects (Baecher, 1987). By requiring that a design meet or exceed the reliability typically achieved in practice, the adequacy of a proposed design and the field data supporting the design can be assessed. In some cases more field data will be required to reduce the uncertainty in the input parameters to the design model and thereby improve the reliability of the design. In other cases, this analysis will reveal that additional sampling will not produce the designed reliability and the system must be redesigned (Graettinger and Dowding, 2000).

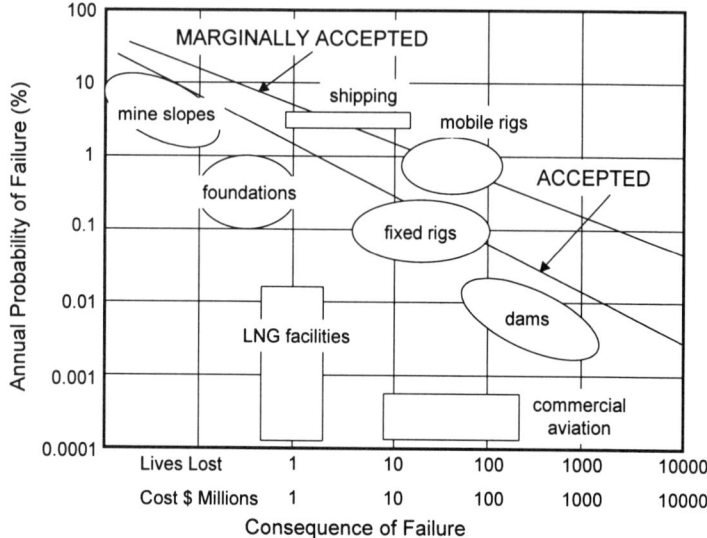

Fig. 9. Empirical failure rates for various engineering projects (Baecher, 1987).

REFERENCES

Baecher, G. B., and Ingra, T. S. (1981). "Stochastic FEM in Settlement Predictions" J. Geotech. Engrg. Div., ASCE, Vol. 107, No. GT4, Proc. Paper 16179, pp. 449-463.

Baecher, G. B. (1987), Geotechnical Risk Analysis User Guide, FHWA/RD-87-011, Fed.Hwy.Admin.,McLean, p.55

Ditmars, J. D., Baecher, G. B., Edgar, D. E., and Dowding, C. H. (1988). "Radioactive Waste Isolation in Salt: A Method for Evaluating the Effectiveness of Site Characterization Measurements." U. S. Department of Energy, Office of Civilian Radioactive Waste Management Salt Repository Project Office, Available from National Technical Information Service, U. S. Dept. of Commerce, Springfield, VA.

Finno, R. J. (1996). "LIGO Report." Internal Northwestern University Document, Civil Engineering Department Northwestern University, Evanston, IL.

Graettinger, A. J. (1998). Reliability-Based Exploration: A Quantitative Method for Evaluating and Directing Subsurface Characterization, Ph.D. Thesis, Northwestern University, Evanston IL.

Graettinger, A. J. and Dowding, C. H. (1999). "Directing Exploration with 3D FEM Sensitivity and Data Uncertainty." Journal of Geotechnical and Geoenvironmental Engineering, ASCE, Vol. 125, No. 11, pp. 959-967.

Graettinger, A. J. and Dowding, C. H. (2000). "Quantifying the Sufficiency of Exploration While Accommodating Qualitative Judgement." Journal of Mathematical Geology, IAMG, Accepted December 1999.

Harr, M. E. (1996). Reliability-Based Design in Civil Engineering, Dover Publications, Inc., New York.

Isaaks, E. H. and Srivastava, R. M. (1989). An Introduction to Applied Geostatistics, Oxford University Press, New York, pp. 278-322.

Lacasse, S. and Nadim, F. (1996) "Uncertainties in Characterizing Soil Properties", Uncertainty in the Geologic Environment: From Theory to Practice, ASCE Geotech. Spec. Publ. No. 58, ASCE, New York, NY, pp. 49-75.

Reeves, H. W., Lee, J., Dowding, C. H., and Graettinger, A. J.(1999). "Reliability-Based Evaluation of Groundwater Remediation Strategies", ModelCARE'99 Conference Proceedings, Institute of Hydromechanics and Water Resources Management, International Conference on Calibration and Reliability in Groundwater Modeling Coping with uncertainty, Zurich, Switzerland.

Ripley, B. D. (1987). Stochastic Simulation, John Wiley & Sons, New York, pp. 237.

Tomasko, D., Reeves, M., and Kelley, V. A. (1987). "Parameter Sensitivity and Importance for Radionuclide Transport in Double-Porosity Systems." Proc. of the Conference on Geostatistical, Sensitivity, and Uncertainty Methods for Ground-Water Flow and Radionuclide Transport Modeling, 87 DOE AECL, pp. 297-321.

Townley, L. R. and Wilson, J. L. (1985) "Computationally Efficient Algorithms for Parameter Estimation and Uncertainty Propagation in Numerical Models of Groundwater Flow", Water Resource Research, Vol. 21, No. 12, pp.1851-1860.

DATABASE OF IN SITU TEST RESULTS FROM RECIFE SOFT CLAYS

By Roberto Q. Coutinho[1] M. ASCE, Joaquim T. R. Oliveira[2], and Leonardo M. Santos[3]

ABSTRACT

This paper presents the database of in situ test results (SPT, CPTU, PMT, DMT and Field Vane) from Recife soft clays. The geotechnical database was developed using tools of the information technology and presents all in situ tests results from two research sites. The database has shown to be a very good tool in research and projects, and it has also been used to help the geotechnical education of students in graduate and undergraduate geotechnical engineering courses.

INTRODUCTION

Throughout the past years, the Geotechnical Group from the Civil Engineering Department of the Federal University of Pernambuco - Brazil (UFPE) has been developing geotechnical studies from laboratory and in situ tests in the Recife lowland soils. The database of laboratory test results was published by Coutinho et al. (1998). This paper will present the database of the in situ test results, its structure, main features, and applications.

The main purposes of the database are: to allow a faster and efficient way for advanced research and statistical analysis of the geotechnical parameters obtained by in situ tests; to give basic information about the equipments used and about the in situ tests performed by the UFPE team; to contribute for spreading the use of others in situ tests (different from SPT) in Brazil, especially in the Northeastern Area; to make easier and more efficient the evaluation/calibration of empirical correlation from literature to the Recife soft clays; to be used as a pedagogical tool to help teaching in graduate and undergraduate geotechnical engineering courses.

---

[1] Associate Professor, Federal University of Pernambuco, Dept. of Civil Engineering, Recife – PE
[2] Civil Eng., Federal University of Pernambuco/Ph.D. Student COPPE/UFRJ-UFPE (Joint Research)
[3] Graduate Student, Federal University of Pernambuco

LOWLAND DESCRIPTION

Recife is situated on the Northeastern coast of Brazil (Figure 1) and presents a plain area formed in the Quaternary Period with influence of salt and fresh water. Soft clay and organic soil deposits are found in about fifty percent of the lowland area, formed in the Holocene Period, with a maximum age about 10,000 years. The land level is close to the sea level. In general, the soft soil deposits are almost entirely below the water table level. The zoom in Figure 1 shows the Recife map with the location of the investigated sites included in the laboratory (Coutinho et al., 1998) and in situ tests databases. The Figures 2 and 3 show the SPT profile and results of characterization tests of the two Recife research sites (RRS).

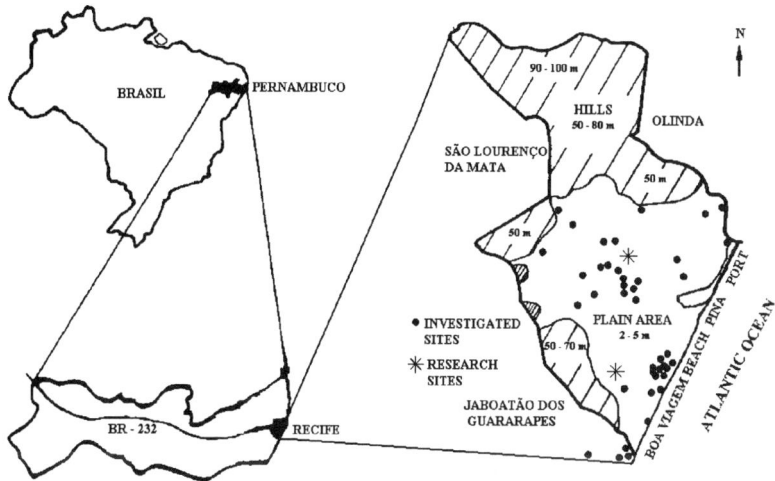

Fig. 1. Location of Recife and Investigated Sites

DATABASE PURPOSES

The Geotechnical Group from the Civil Engineering Department of the Federal University of Pernambuco has been developing a digital database of in situ test results from Recife soft clays since 1997. Firstly, this project was created just to organize all in situ test results and make the research faster and more efficient. Another important goal was to save space required to file a lot of test results. This paper presents the newer version of the database where its purposes and resources have been widely extended. The main purposes of the database now are:

   a) to allow a faster and efficient way for advanced research and statistical analysis of the geotechnical parameters obtained by the in situ tests;

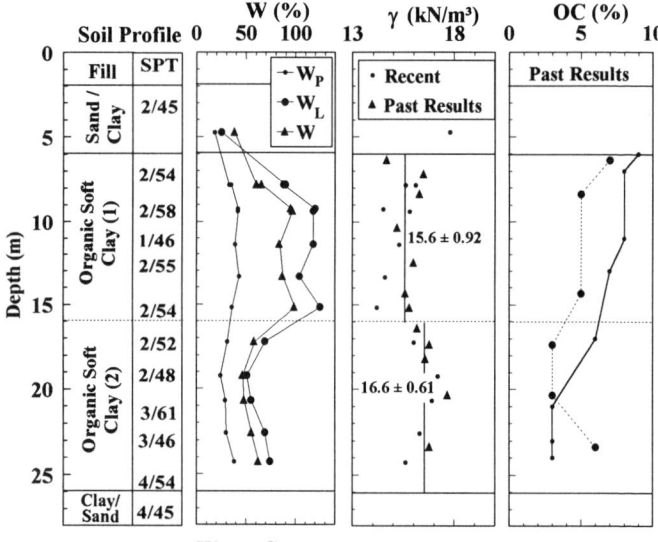

Fig. 2. Results of Characterization Tests - RRS 1 (Coutinho & Oliveira, 1997)

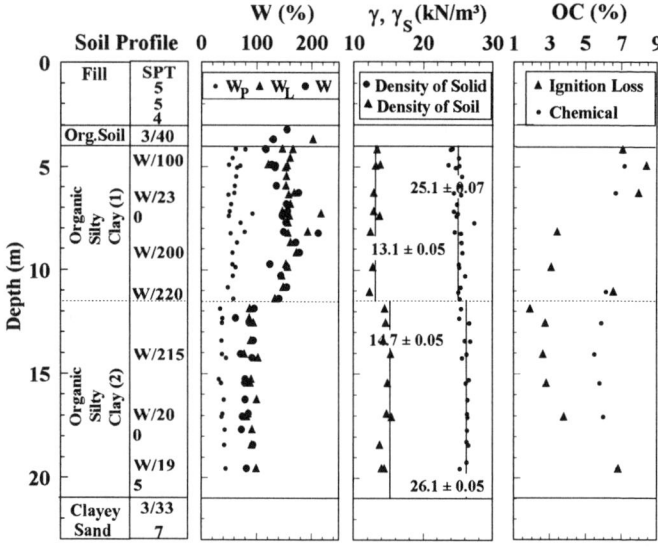

Fig. 3. Results of Characterization Tests - RRS 2 (Coutinho et al., 1998)

b) to give basic information about the equipments used and about the in situ tests performed by the UFPE team;
c) to contribute for spreading the use of others in situ tests (different of SPT) in Brazil, especially in the Northeastern area;
d) to make easier the evaluation/calibration of empirical correlation from literature (CPTU, DMT, PMT and Field Vane) to be used in the Recife soft clays;
e) to join basic information about all the in situ tests (methodology, procedures, equipments, special cares, a summary of geotechnical parameters obtained and practical applications) to be used as a pedagogical tool to help teaching in graduate and undergraduate geotechnical engineering courses.

The newer version of the database uses the most recent tools of the information technology to achieve these purposes and it could be accessed by either specific software in CD-ROM or by Internet (published data). The access by Internet could be done in a specific web site (there is no URL yet) where the user will be able to get the same information presented in the CD-ROM. The access by Internet presents the advantage of researching from anywhere where there is a computer connected to the Internet. The module of accessing the database by Internet will not be presented here. This paper will be focused in the specific software temporarily labeled as IN SITU 2000 and applications obtained from the database.

THE SOFTWARE IN SITU 2000

The software temporarily labeled as IN SITU 2000 is being developed in the programming language Microsoft Visual Basic 5.0® and works with different relational databases like Microsoft Access®, SQL server® and Oracle®. The IN SITU 2000 is Windows® 9x compatible and presents a graphical interface clean, objective and very intuitive. The minimum system requirements for a good software performance are: Pentium® processor (or equivalent) 100 MHz, free disk space about 30 MB, 32 MB RAM, multimedia kit and operating system Windows® 9x.

The goal is to develop the IN SITU 2000 completely customizable and adaptable to other places and different in situ tests. In this way, researchers or designers from different places could enter specific data (like maps, geographic coordinates, test results) and use the software to analyze it. In a close future, the exchange between researches from different places may build a great database composed of different databases. To make it possible, the software will be available in three different languages: Portuguese, English and Spanish.

The software is composed of three main modules. The first one presents multimedia tools and functions to find and locate the test sites in the city maps. The second part shows all test results in tables, charts and figures showing the site studied, the depth and the type of tests performed, and the analysis of results. There are research tools such as filtering making possible advanced research (combining several criteria by logical operators), data evaluation, and also creation, edition, and printing of reports. The geotechnical parameters obtained, the comparison to laboratory results, the statistical analysis and the practical applications are also present in that part. The third part has pedagogical purposes and it is an innovation in the Northeast Area of Brazil. This part contains a description of each equipment, test procedures, special cares, and a summary of geotechnical parameters obtained and practical applications. It has been used to help the geotechnical education of students in graduate and undergraduate geotechnical engineering courses. The following item will describe each of these main parts using an illustrative example.

ILLUSTRATIVE EXAMPLE

The main software window is showed in Figure 4 where can be seen the Recife city map. From this window the user can interact and use all available tools to research, to evaluate statistical analysis, to study about the tests and so on. The next sequence shows a step-by-step research about the Pressuremeter Test. From the main menu **'Functions'** (Figure 4) the user can load the test results from any in situ test clicking in the submenu **'Load…'** and choosing the specific test, PMT test in this case. IN SITU 2000 will mark all sites and holes on the map where tests were performed. This situation is showed in Figure 5, where the city map was zoomed by the command zoom to show the details. The user can also filter the data just choosing the option **'Filter…'** in the main menu **'Functions'** rather than **'Load…'** (see Figure 4).

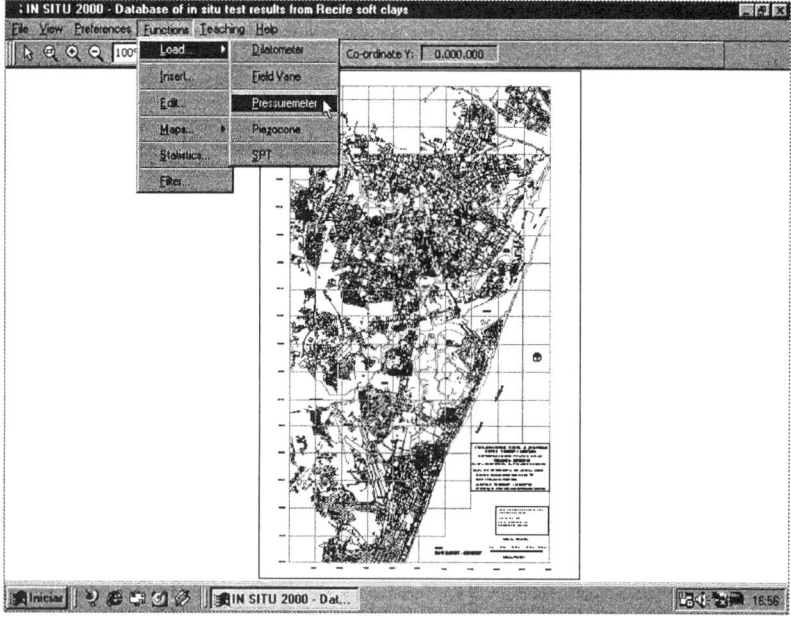

Fig. 4. Main Window of the IN SITU 2000

The filled circles over the map (Figure 5) show the holes where PMT tests were performed while cross marks show holes where another kind of in situ test different from PMT (such as SPT) was performed. The user can configure the marks and change shapes, colors and size. From this point, the user is able to know all available results from each hole just clicking the mouse right button over one of the filled circles so a popup menu will appear showing the in situ test information available to the hole. In the example showed in Figure 5 the hole label is 'PMT: S1F2' and the available information of the results are tables and charts. The option 'Show Figures...' is disabled because there are no figures (Photos, etc.) about this hole. The geographic co-ordinates of the hole are also available as can be seen in the toolbar (Figure 5) just moving the mouse over the map.

After clicking the option 'Show Charts...' (Figure 5) the software will show a window like that presented in Figure 6 where is showed a brief description about the site and the borehole (address, borehole label, co-ordinates, etc.).

In Figure 6 it also can be seen some information about the soil profile and charts from the test results. The user can get the other charts available just clicking on the 'Next' button. Specific values can be obtained moving the mouse over the charts (Figure 6). Another database improvement was the function 'Filter...' (see Figure 4). Now the user can execute advanced research combining several criteria by logical operators (see Figure 7).

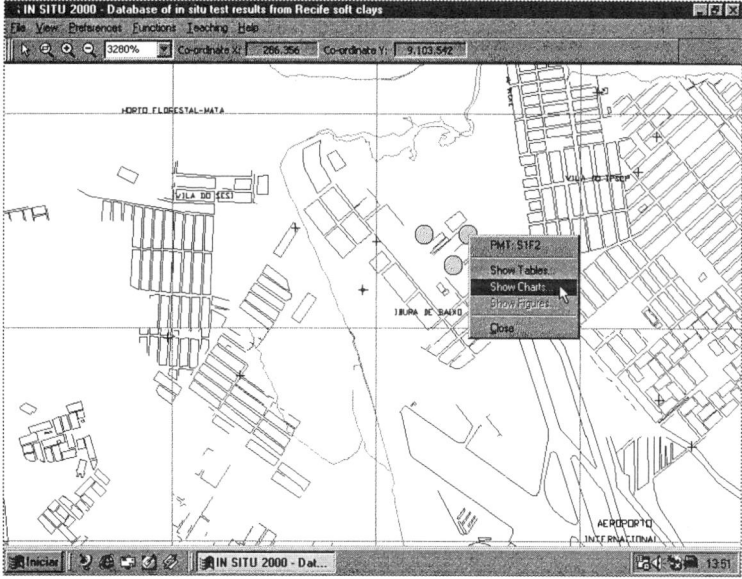

Fig. 5. PMT Tests Performed Located in the City's Map.

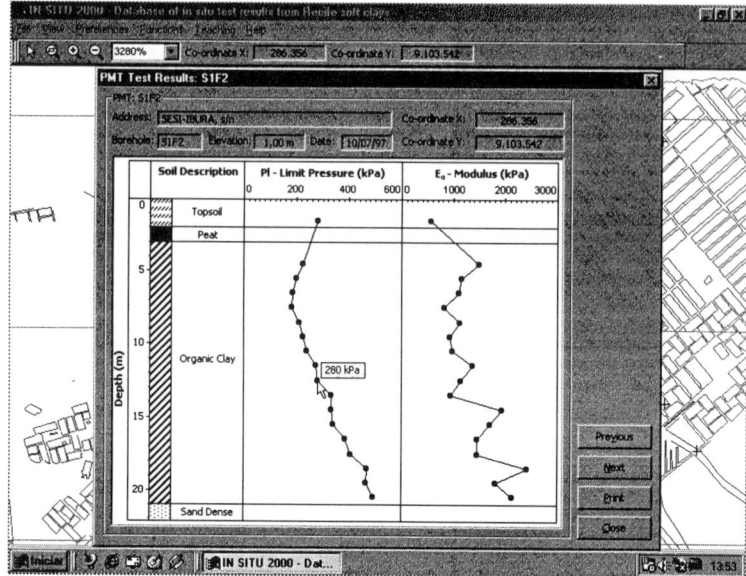

Fig. 6. Available Information: PMT Charts

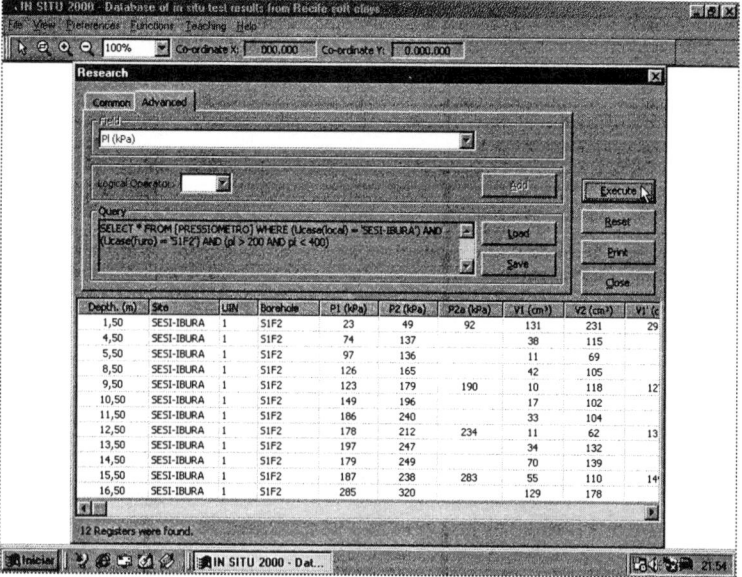

Fig. 7. Advanced Research (Example).

GEOTECHNICAL EDUCATION

The Geotechnical Group of the UFPE has been working recently in the improvement of geotechnical engineering teaching. In fact, the third part of the IN SITU 2000 was created for pedagogical purposes to help the geotechnical education of students in graduate and undergraduate courses. It represents an innovation in the Northeastern Brazil and probably the first experience in Brazil. The third module of the IN SITU 2000 is based on HTML and was developed using the software Microsoft® HTML Help Workshop. It means that the user will use this module as the Internet. It also makes possible the self-learning of some geotechnical topics. In Figure 8 can be seen an example contents of this part. It contains a description of each equipment, theory, test procedures, special cares, and a summary of geotechnical parameters obtained and practical applications. This part presents search tools to find specific subjects or words. There is also an item containing a list of important references about the in situ tests including a full version of all recent published papers by the UFPE Team.

SUMMARY OF GEOTECHNICAL PARAMETERS OBTAINED AND PRACTICAL APPLICATIONS

In this section it will be presented examples profiles with some important geotechnical parameters in the Recife soft clays from the in situ tests investigated (see Coutinho & Oliveira, 1997; Coutinho et al., 1999). These results are all included in the In Situ 2000

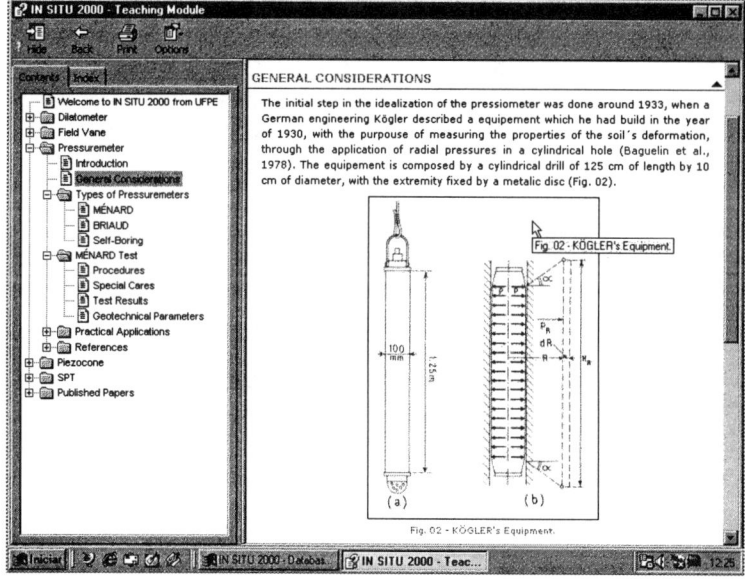

Fig. 8. Teaching Module of IN SITU 2000 (Example).

database. A brief comparison with laboratory results will be made. The correlations were obtained using the software module that allows data evaluation and statistical analysis. The comparison between laboratory and in situ test results was done crossing data from the laboratory and in situ database versions.

Overconsolidation Ratio (OCR)

Sully et al. (1988) presented a correlation to a piezocone with two measurements of pore pressure (u_1 and u_2):

$$OCR = 0.49 + 1.5PPD, \text{ where: PPD = pore pressure difference } (u_1 - u_2) \qquad (1)$$

Figure 9b shows an example of OCR calculated from this CPTU correlation and results of oedometer tests from RRS 2 (SESI-Ibura). The values of piezocone are slightly higher than values of laboratory tests. In the case of piezocone with just one pore pressure measurement the results from the correlations by Lunne et al. (1989) and Kulhawy and Mayne (1990) were presented by Coutinho et al. (1999).

From the dilatometer tests OCR values which present the best agreement with oedometer data are those calculated using the Lunne et al. (1989) correlation (see Figure 9c):

$$OCR = 0.3K_D^{1.17} \qquad (2)$$

Using results from Field Vane Tests, it is possible to obtain good results of OCR values from the Mayne & Mitchell (1988) equation (Figure 9b):

$$OCR = \alpha_{FV}\left(\frac{S_U}{\sigma'_{v0}}\right)_{FV}, \text{ where } \alpha_{FV} = 22IP^{-0.48} \qquad (3)$$

Coefficient of earth pressure at rest (k_0)

In situ horizontal stress, σ'_{ho} (or coefficient of earth pressure at rest, k_0) is an important and very difficult geotechnical parameter to get good measurement with any device. In general, there is an uncertain reliability because of the scarcity of referent values (Lunne et al., 1990). Values of k_0 were obtained from PMT (Cavalcante, 1997) and using correlations from dilatometer and piezocone tests. The laboratory correlation (Mayne and Kulhawy,1982) was used like a reference. The following equations presented good agreement:

a) Piezocone:

$$k_0 = 0.5 + 0.11\frac{(u_1 - u_2)}{\sigma'_{vo}} \qquad \text{(Sully and Campanella, 1991)} \quad (4)$$

b) Dilatometer:

$$k_0 = 0.34 K_D^{0.54}$$ (Lunne et al., 1990) (5)

The Figure 9d presents the average values of k_0 obtained by using these correlations from SESI-Ibura, showing that DMT results were very close to the "laboratory" correlation results. The k_0 - values from PMT are smaller than those from laboratory and the scatter of data is high.

Undrained shear strength (S_U)

The shear strength parameter in undrained condition (S_U) was obtained from CPTU, vane, PMT, and DMT tests. The S_U - in situ tests empirical correlation that were used are referred to the triaxial compression test or to the field vane test depending on the case. The cone resistance is related to undrained shear strength by the cone factor:

$$N_{KT} = \frac{(q_T - \sigma_{v0})}{S_U}$$ (6)

Despite theoretical advances, empirical correlation is still the most commonly used for the interpretation of piezocone tests. In this study, the empirical correlation proposal by Tavenas & Leroueil (1987) - N_{KT} vs. IP (field vane tests) and the correlation proposal by Lunne et al. (1985) - N_{KT} vs. B_q (compression triaxial tests) were used.

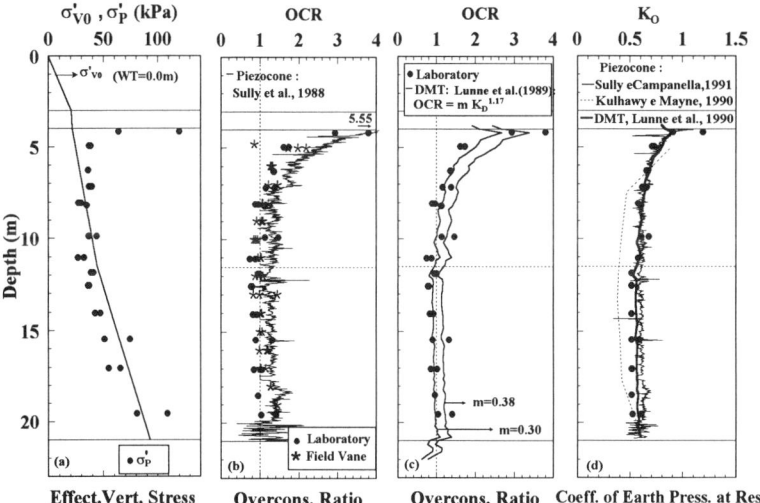

Fig. 9. Stress History and In Situ Horizontal Stress Parameters - RRS 2 (Coutinho et al., 1999)

The interpretation of dilatometer results was done in this study using the empirical correlations proposal by Marchetti (1980), by Lacasse & Lunne (1988) (see also Lunne et al., 1989). Lacasse and Lunne correlations for field vane test (8) and Laboratory triaxial compression test (9):

$$S_U = A\sigma'_{v0}(0.5K_D)^{1.25}; \text{ where A: } 0.17\text{-}0.21 \tag{7}$$

$$S_U = 0.20\sigma'_{v0}(0.5K_D)^{1.25} \tag{8}$$

They are very similar to the Marchetti correlation (9) and for limited space conditions the results from only one correlation will be presented in each part of Figure 10.

$$S_U = 0.22\sigma'_{v0}(0.5K_D)^{1.25} \tag{9}$$

$$S_U = \frac{P_L^*}{7.8}, \text{ where } P_L^* = \text{corrected limit pressure} \tag{10}$$

Figures 10 shows S_U values obtained in triaxial compression tests and Field Vane tests as well as mean values obtained from the piezocone, pressuremeter, and dilatometer. In general, the results obtained in this paper are in agreement with other studies (Lunne et al. 1989; Marchetti, 1997) confirming the possibility of a good prediction of S_U from the piezocone and Marchetti dilatometer tests. For the Recife soft clays Powell (1990) correlation from PMT also seems to present good results. Table 1 presents a summary of the comparative study of geotechnical parameters in Recife soft clays from laboratory and in situ tests.

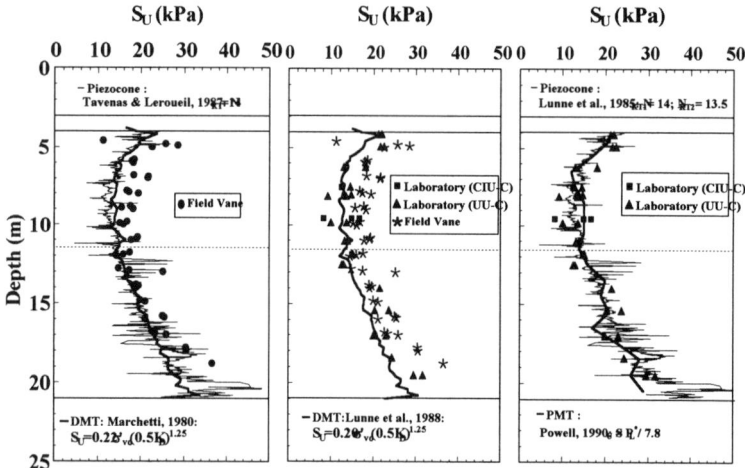

Fig.10. Su from Laboratory and In Situ Tests - RRS 2 (Coutinho, et al., 1999)

Table 1. Comparative Study Laboratory x In Situ Tests.

Parameter	CPTU	In Situ Test DMT	PMT	FVT
OCR	Sully et al. (slightly higher)	Lunne et al. (1989) (close)		Mayne & Mitchell (1988) (close)
k_0	Sully & Campanella (1991) (close)	Lunne et al. (1990) (very close)	High scatter (smaller)	
$S_{U,LAB}$ *	Lunne et al.(1985) (close)	Marchetti (1980) (slightly higher/close)	Powell (1990) (close)	Close
$S_{U,vane}$ **	Tavenas & Leroueil (1987) (slightly smaller)	Marchetti (1980) (slightly smaller)		

Note: * - Triaxial UU-C and CIU-C; ** - only one research site

CONCLUSIONS

The database of in situ test results from Recife soft clays has shown to be a good tool to be used in teaching, research and practical applications. The newer version presented here will increase this potential. The comparison between the two databases results (laboratory and in situ tests) from Recife soft clays permit to obtain the in situ correlation that give the most closely matching laboratory data. This is an important information to the regional geotechnical community and practical applications.

ACKNOWLEDGMENTS

The authors are grateful to the CNPq - Brazilian Research Council for the financial support given to the research project and also to the undergraduate student João Carlos Cunha that provided valuable help in the preparation of this paper.

REFERENCES

Cavalcante, E. H. (1997) One contribution to the stress - strain behavior study of the Recife soft clay from Ménard Pressuremeter - M Sc. thesis - UFPB - Brazil (in Portuguese).

Coutinho, R. Q. & Oliveira, J. T. R. (1997) Geotechnical characterization of a Recife soft clay - laboratory and in situ tests - 14th ICSMGE, 1: 69-72, Hamburg, Germany.

Coutinho, R. Q., J. T. R. Oliveira & Oliveira, A. T. J. (1998) Geotechnical Site Characterization of Recife Soft Clays. 1st International Symposium on Site Characterization, 2: 1001-1006, Atlanta, USA.

Coutinho, R. Q., J. T. R. Oliveira, Pereira, A.C. & Oliveira, A. T. J. (1999) Geotechnical Characterization of a Recife Very Soft Organic Clay - Research Site 2. XI Panamerican Conference on Soil Mechanics and Geotechnical Engineering - Vol. 1 : 275-282- Foz do Iguaçú – Brazil.

Kulhawy, F. & Mayne, P. (1990) Manual on estimating soil properties for foundation design, Report N.EL-6800 Electric Power Research Institute, Cornell University, Ithaca, N.4, 250pp.

Lacasse, S. & Lunne, T. (1988) Calibration of dilatometer correlations. *Proc. ISOPT- 1*, Orlando, Fl., USA Vol.1: 539-548.

Lunne, T., H. P. Christoffersen & T. I. Tjelta (1985) Engineering use of piezocone data in north sea clays, *Proc. of 11th ICSMFE*, San Francisco, 2: 907-912.

Lunne, T., S. Lacasse & Rad, N. S. (1989) Pressuremeter testing and recent developments - part I: All tests except SPT, General Report, session 2. *Proc. 12th ICSMFE*, Rio de Janeiro, 4: 2339-2403.

Lunne, T., J. J. M. Powell, E. A. Hange, I. M. Uglow & Mokkelbost, K. H. (1990) Correlation of dilatometer readings to lateral stress. Specially session on measurement of lateral stress. 69th Annual Meeting of the Transportation Research Board, Washington, D.C., USA

Marchetti, S. (1980) In situ tests by flat dilatometer - ASCE Jul, GE, 106(3): 299-321.

Marchetti, S. (1997) The flat dilatometer design applications - Keynote Lecture - 3rd Geotech. Eng. Conf. Cairo Univ. - 26pp, Jan - Cairo.

Mayne, P.W. & Kulhawy, F. H. (1982) K_0 - OCR relationship in soil, ASCE - JGED, 108(GT6): 851-872.

Mayne, P.W. & Mitchell, J.K (1988) Profiling of overconsolidation ratio in clays by field vane. *Canadian Geotechnical Journal*, Vol. 25: 150-157.

Powell, J. J. M. (1990) A Comparison of Four Different Pressuremeters and their Methods of Interpretation in a Stiff Heavilly Overconsolidated Clay. *Proc. 3rd Int. Symp. of Pressuremeters*, Oxford: 287-298.

Sully, J. P., Campanella, R. G. & Robertson, P.K. (1988) Interpretation of Penetration Pore Pressure to Evaluate Stress History in Clays. *Proc. of the ISOPT - 1*, Orlando, Florida, USA, 2: 993-999.

Sully, J. P. & Campanella, R. G. (1991) Effect of Lateral Stress on CPT Penetration Pore Pressure. Journal of Geotechnical Engineering, ASCE, 117(7): 1082-1088.

Tavenas, F. & Leroueil, S. (1987) Laboratory and In Situ Stress-Strain-Time Behavior of Soft Clays: A State-of-the-Art. Proc. of the Intern. Symp. on Geotech. Eng. of Soft Soils, 2: 3-48, Mexico City, Mexico.

GEOSTATISTICAL SAMPLE LOCATION SELECTION IN EXPEDITED HAZARDOUS WASTE SITE CHARACTERIZATION

Greg A. Stenback[1] and Bruce H. Kjartanson[2] M. ASCE

ABSTRACT

A geostatistical sequential sample location selection method is developed for Expedited Site Characterization or accelerated site investigations. The method provides a measure of the uncertainty regarding contaminant concentrations at unsampled locations on the basis of the existing contaminant concentration data. Sample locations are selected within regions where the probability that contamination exceeds a specified threshold indicates a high level of uncertainty. Samples are spaced according to the range of influence determined from a variogram developed from measured contaminant concentrations. Sample collection ceases when the contaminant plume is bounded by a probability contour indicating low probability that a contaminant concentration will exceed the threshold at unsampled locations outside the bounded region. The method is demonstrated with trichloroethylene data collected during an Expedited Site Characterization project at the U.S. Department of Energy Savannah River Site in 1995.

INTRODUCTION

The questions of where to take subsequent samples and when to stop sampling usually arise during a hazardous waste site characterization. The U.S. Environmental Protection Agency Data Quality Objectives (DQO) process (EPA, 1993) suggests data collection strategies and stopping rules based on the use to which the data will be put and statistical hypothesis tests for reaching specified conclusions from the data. The DQO statistical methods and statistical tests described in EPA (1989, 1992a, and 1993) generally assume that probabilistic sampling methods, e.g. simple random, stratified random, or systematic sampling, are employed. EPA (1992b) discusses some statistical methods for analysis of groundwater data where serial or spatial correlation are likely. Contaminant concentration data often display spatial correlation that may be

---

[1] Post-Doctoral Research Associate, Dept. of Civil and Construction Engineering, Iowa State University, Town Engineering Building, Ames, Iowa, 50011
[2] Associate Professor, Dept. of Civil and Construction Engineering, Iowa State University, Town Engineering Building, Ames, Iowa, 50011

characterized using statistical methods for spatial data. In such cases the locations and magnitudes of existing data can be used to suggest locations for subsequent sampling and to help the site manager decide when to stop sampling and conclude that sufficient data exist to adequately characterize the site. David and Yoo (1993), Englund and Heravi (1994), Johnson (1993 and 1996), Stenback (1996), Van Tooren and Mosselman (1997) and workers at Sandia National Laboratory (1998) have developed methods to utilize information contained in data as it is collected to aid this decision making process. A more general description of geostatistical methods and terminology may be found in Isaaks and Srivastava (1989) and Cressie (1991).

David and Yoo (1993) developed an adaptive sample location selection algorithm that appears to produce a nearly uniform sample spacing over the region of interest. The method uses spatial interpolation but does not require a spatial correlation model. Accordingly, the method does not account for observed spatial correlation. They suggest a stopping rule based on the convergence of a discrepancy function which estimates the volume, or average distance, between a map based on all the currently available data, and a map based on one half of the current data.

Englund and Heravi (1994) have incorporated a cost function into an adaptive sampling strategy that uses block kriging to estimate average contaminant concentrations over remedial units. The method is developed for soil remediation and seeks to minimize the expected total characterization and remediation costs by effective placement of future samples based on currently available data. The major difficulty with this method is the development of a realistic cost structure to adequately characterize intangible costs to society, such as negatively impacted public health or ecosystems, associated with failure to remediate contaminated soil.

Johnson (1993 and 1996) has developed an adaptive sampling technique using Bayesian analysis to quantitatively integrate "soft" information (e.g. historical information, geophysical survey results, preliminary modeling results, or personal experience) with quantitative analytical data. New sample locations are chosen to maximize the area declared clean, maximize the area declared contaminated, or minimize the area declared as state uncertain. This method is very computationally intensive and is sensitive to initial contaminant spatial distribution assumptions.

Van Tooren and Mosselman (1997) have proposed that kriging be used to develop site maps showing the probability that unsampled regions have contaminant concentrations exceeding a specified threshold, such as an action level or cleanup level. They suggest that future samples be located in those regions with high likelihood of misclassifying contaminated areas as clean or uncontaminated areas as needing remediation; these regions have probabilities between 0.4 and 0.6 on the probability maps.

Sandia National Laboratory (1998) has developed a probabilistic approach called Smart Sampling™ that involves an initial collection of site soil contaminant concentration data followed by characterization of the concentration distribution and spatial correlation. Geostatistical simulations using the observed data and spatial correlation model are performed to develop many possible models of the contaminant concentration spatial distribution at unsampled locations. Using these many possible models, a probability map can be constructed to show the frequency with which the simulated values exceed a given action or cleanup level. Based on this information, maps can be developed to show the extent of excavation required to clean to a specified level with a specified probability of leaving contaminated soil in place. Cost curves can be developed to show the effect that the choice of cleanup level will have on the remediation cost.

The sample location selection method developed by Stenback (1996) is the subject of this paper. The method is similar to Van Tooren and Mosselman's, as described above, but provides additional guidance on the distance between sample locations and the development of a stopping rule indicating when a sufficient amount of data have been collected. Stenback's method is also similar to the Sandia National Laboratory approach, but bypasses the simulation step by use of a direct calculation.

The problem of deciding where to place subsequent samples and when to stop sampling may be particularly acute during an accelerated site characterization that uses a flexible, dynamic work plan with on-site decision making whereby rapid data integration, visualization, and interpretation are critical. The sample location selection method described in this paper uniquely addresses these issues. Sample location in this context typically refers to the collection of a soil or groundwater sample using either a direct push percussion or cone penetrometer sampling probe. Once a contaminant plume and geologic factors that control contaminant migration are well defined, then an optimal monitoring well network can be developed.

Our geostatistical adaptive sampling method was developed for application within an accelerated site characterization program whereby analytical data, and possibly geologic site characterization data, are rapidly obtained while in the field and all currently available data are used to determine future environmental sample locations. Our method provides input to assist the on-site decision-making process whereby sample locations are selected by the site manager on the basis of all relevant information available including a statistical data analysis and an understanding of the geologic and/or hydrogeologic controls on contaminant migration. An example using concentration measurements for trichloroethylene (TCE) dissolved in groundwater collected during an Expedited Site Characterization project at the U.S. Department of Energy (USDOE) Savannah River Site is used to illustrate the method.

EXPEDITED SITE CHARACTERIZATION

Expedited Site Characterization (ESC), first developed by the USDOE's Argonne National Laboratory (Burton et al. 1993), is a site characterization approach that focuses on utilizing nonintrusive and minimally intrusive investigative techniques to efficiently and thoroughly characterize hazardous waste sites. A fundamental principle of ESC is the ability to perform on-site decision making, within the framework of a dynamic work plan, allowing site characterizations to be completed in a single, phased effort rather than in a protracted, iterative manner. Following an initial organization and planning phase, the site is characterized typically in two phases, each lasting from about two to four weeks in the field. Phase I is focused on the geologic and hydrogeologic characterization of the relevant contaminant migration pathways and characterization of the contaminant source(s), including defining a narrowed list of contaminants of potential concern. Phase II is focused on defining the distribution and concentration of contaminants and further development and refinement of geologic and hydrogeologic controls on migration pathways using direct push technologies, if required. Rapid data collection, interpretation and visualization technologies are used to update the conceptual site model on a daily, or possibly hourly, basis as the investigation proceeds. For further information on the ESC process, the reader may consult Stenback et al. (1998).

Groundwater or soil sampling equipment, such as percussion probe and/or cone penetration test equipment, and a mobile analytical laboratory capable of testing for each contaminant of concern in each media of concern are typically used during an ESC Phase II investigation. Data from the mobile analytical laboratory are generally available within hours of sample collection. The success of the methodology relies on the ability to rapidly integrate and evaluate data to assess uncertainty in the conceptual site model, effectively place subsequent samples, and decide when enough data have been collected. While experience and judgment play a role in deciding where to sample next and when to stop sampling, geostatistical tools can greatly assist the decision-making process.

PROBABILITY BASED GRAPHICAL SAMPLE LOCATION SELECTION

A graphical approach to the selection of sequential sample locations may be developed by defining the regions within the site corresponding to the conditional probability $P(Z(\mathbf{s}) > T \mid \underline{Z})$, where $Z(\mathbf{s})$ is a contaminant concentration at spatial location $\mathbf{s}$ ($\mathbf{s} = (x,y)$ or (x,y,z) spatial coordinates), T is a threshold concentration such as an action level or cleanup standard, and $\underline{Z} = \{Z(\mathbf{s}_1), \cdots, Z(\mathbf{s}_n)\}^T$ is the existing data. The basic premise is that regions where $P(Z(\mathbf{s}) > T \mid \underline{Z})$ is below a level p_1 are most likely "clean" and do not need further sampling and regions where $P(Z(\mathbf{s}) > T \mid \underline{Z})$ is above a level p_2 are most likely contaminated at or above the threshold level and do not need further sampling. The values for p_1 and p_2 are chosen to give acceptable and achievable probability levels, e.g. $p_1 = 0.1$ or 0.2 and $p_2 = 0.8$ or 0.9. Regions with probabilities between p_1 and p_2 have a relatively large uncertainty regarding whether contamination exceeds the threshold or not. It is those areas with greatest uncertainty that should be sampled.

A probability map showing the estimated $P(Z(\mathbf{s}) > T \mid \underline{Z})$ for unsampled locations $\mathbf{s}$ within the site may be constructed using a number of methods including indicator kriging, ordinary kriging, or the method developed by Stein (1992). Indicator kriging is an approximation to the probability $P(Z(\mathbf{s}) > T \mid \underline{I})$, where the conditioning is with respect to the indicator values $\underline{I} = \{I(\mathbf{s}_1), \cdots, I(\mathbf{s}_n)\}^T$ assuming knowledge of the indicator variogram (Cressie, 1991, page 282). The indicator values are defined as $I(\mathbf{s}_i)$ = 1 if $Z(\mathbf{s}_i) > T$ and 0 otherwise. As a nonparametric method, indicator kriging is useful whenever a suitable transformation to a normal distribution cannot be found, and it provides the ability to easily deal with non-detect data (data reported as below the analytical capability for quantification). Assuming the Z process is Gaussian and is well characterized by the variogram model, the probability $P(Z(\mathbf{s}) > T \mid \underline{Z})$ can be estimated using ordinary kriging by inverting a one-sided prediction interval for an observation at an unsampled location. The upper bound for this prediction interval is $\hat{Z}(\mathbf{s}) + z_\alpha \sigma_k(\mathbf{s})$, (Cressie, 1991, page 155). Starting with this upper bound, we have $P(Z(\mathbf{s}) > \hat{Z}(\mathbf{s}) + z_\alpha \sigma_k(\mathbf{s}) \mid \underline{Z}) = 1 - \Phi(z_\alpha)$, where $\hat{Z}(\mathbf{s})$ is the kriging predicted value, $\sigma_k(\mathbf{s})$ is the kriging prediction standard deviation, and $\Phi(\cdot)$ is the standard normal cumulative distribution function. Equating the upper bound with the threshold T and noting that $1 - \Phi(z_\alpha) = \Phi(-z_\alpha)$ gives

$$P(Z(\mathbf{s}) > T \mid \underline{Z}) = \Phi\left(\frac{\hat{Z}(\mathbf{s}) - T}{\sigma_k(\mathbf{s})}\right). \tag{1}$$

Stein's method was developed for spatial prediction with partially truncated, normally distributed data and provides a direct estimate of $P(Z(\mathbf{s}) > T \mid \underline{Z})$, as well as a predicted value and prediction variance. The method is applicable to data where the truncation, or censoring, is due to an analytical method detection limit resulting in some non-detect data. A computer simulation is used to generate values from the joint distribution of the existing data and the result at an unsampled location conditioned on censored data; see Stein (1992) for additional details. In the absence of non-detect data, Stein's method and ordinary kriging will produce the same results, but ordinary kriging is much faster. This choice of methodologies provides the practitioner flexibility to properly address statistical assumptions and non-detect data within the framework of our methodology.

Sample locations are selected from within the region where $P(Z(\mathbf{s}) > T \mid \underline{Z})$ is between p_1 and p_2 in such a way that they are not too close to existing samples, where the measure of closeness is based on the range of influence as characterized by the variogram. To avoid partially redundant data and yet stay within the range of spatial correlation, a sample spacing near one half the range of influence is generally adequate. However, to cover a larger area with fewer samples, a sample spacing nearer the range of influence might be used. Sampling continues until the plume is surrounded by a low probability contour, say 0.1, indicating that the plume boundary (at the specified threshold) is assumed to be adequately defined and further sampling is not necessary. While we might desire a very low probability boundary, say 0.05, to provide high assurance that the plume boundary is well established, our experience indicates that excessive sampling might be required to achieve this level of assurance. This method does not output specific map coordinates where samples should be collected, but rather, it indicates regions of high uncertainty from which samples might be collected to reduce that uncertainty. The actual sample locations are chosen by the site manager.

The method is ideally suited to Phase II of an ESC project in that it can be quickly applied in the field, provides a frequently updated graphical representation of the contaminant distribution, and provides a simple stopping rule. Geologic and hydrogeologic factors are used qualitatively to help select sample locations (e.g. downgradient of the source), and sample depths. The following sections of this paper provide an example application of this methodology; to focus on the geostatistical method, some details of the actual site characterization have been omitted.

OIL SEEPAGE BASIN SITE AND ESC PROGRAM

The principles of ESC were incorporated into a unit assessment activity carried out within the RCRA/CERCLA Facility Investigation/Remedial Investigation (RFI/RI) at the D-Area Oil Seepage Basin at the USDOE's Savannah River Site (SRS), South Carolina in the summer of 1995 (Stenback et al., 1998). The SRS occupies an area of approximately 775 square kilometers within the Atlantic Coastal Plain. The oil seepage basin site is situated near the western border of the SRS on alluvial deposits of the Savannah River. The immediate waste site, occupying an area of about 75 m by 135 m (Figure 1), comprises several unlined trenches into which used oil unsuitable for burning, waste solvents, and other debris were deposited. Historical records indicate that the maximum depth of the trenches was about 2.5 m. Periodic burndown of the trench contents was performed to make room for additional material. In 1975, after about 23 years of operation, the trenches were taken out of service and filled in with soil.

Several sampling events and preliminary investigations conducted at the site prior to this ESC effort indicated the presence of metals, chlorinated solvents and pesticides in the site soils and trench fill. Four water table aquifer monitoring wells and six piezometers (see Figure 1) were installed to a maximum depth of 12.8 m prior to the initiation of the ESC work. Sampling and analysis of groundwater from the monitoring wells revealed the presence of chlorinated solvents above their respective Maximum Contaminant Levels (MCLs) for drinking water. Water level measurements from the piezometers established a west-southwesterly direction of ground water flow under a very low gradient ranging from 0.0005 to 0.003. Depth to the water table ranged from 1.2 m to 5.5 m below ground surface over the course of ten years of measurements. Limited geologic information available prior to the ESC work indicated that the near-surface soils are primarily fine to medium fluvial quartz sands along with silty and clayey sands and intermittent silt/clay lenses. The depth and continuity of a regional aquitard, referred to as the Gordon Confining Unit or "green clay", was unknown at this site but was expected to occur at a depth of less than 20 m.

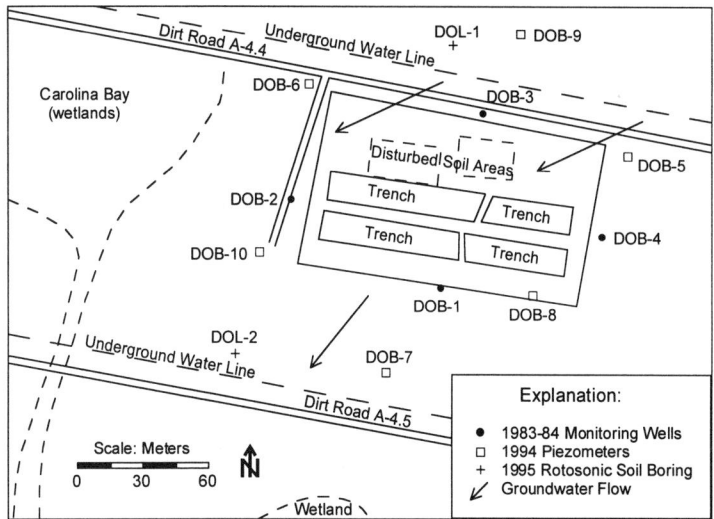

Figure 1. Savannah River Site oil seepage basin.

The site was characterized in two ESC field phases; both phases, with an intermediate meeting with regulators to negotiate target contaminants, were completed in a period of about 3 months. Phase I focused on defining the geology and hydrogeology of the migration pathways by drilling two boreholes (DOL-1 and DOL-2, Figure 1) into the green clay aquitard using rotosonic coring techniques and using borehole and surface geophysical methods (GPR and electrical conductivity). Two weak aquitards, composed of sandy units with elevated silt and clay contents and up to about 1.5 m thick, were identified at depths of about 3 m and 7.6 m. In addition, the green clay aquitard was located at depths from about 11 to 14 m. A list of site specific contaminants (SSCs), used to focus the analytical efforts for Phase II, was developed from analysis of 14 groundwater samples collected during Phase I from the locations shown on Figure 4. The SSC list included trichloroethylene (TCE), tetrachloroethylene, vinyl chloride, one pesticide (α-BHC) and four metals (antimony, arsenic, beryllium and manganese).

Phase II involved defining the geology and hydrogeology of migration pathways using location specific direct push technologies. This activity focused on confirming the continuity and defining the configuration of the weak aquitards and the green clay. This was followed by groundwater sampling and on site analyses to track the SSC plumes. Groundwater samples were collected from multiple depths at 73 map locations (shown in Figure 7) using the site stratigraphic model as a guide, to ensure that contamination, if present, was detected. Close to 200 samples were collected and analyzed in 12 days. Percussion probing screen point or mill-slotted equipment was used to collect the depth discrete samples. Analytical measurements were made on-site using EPA SW-846 methods with definitive data type (EPA, 1993) QA/QC procedures. Although eight SSCs were tracked during the Phase II ESC investigation, only the TCE data are used in this paper to illustrate the geostatistical tools and analyses.

GEOSTATISTICAL ANALYSIS AND SAMPLE SELECTION: TCE PLUME

At each sample location, ESC Phase II groundwater samples were collected from up to seven depths using the geologic/hydrogeologic model and nearby sample results to guide sample location. Although the plume is three dimensional (and possibly variable with time), we examine a projection of the plume in a two dimensional plan view by modeling the maximum value observed from the depth discrete samples collected at the 73 sample locations. Modeling the plume in this manner was reasonable at this site because the discontinuous nature of the silty and clayey lenses within the alluvial soils allowed for mixing of groundwater across depths. Figure 2 shows a normal probability plot of these TCE concentrations. The normal probability plot indicates that, although about 59 percent of the data are below the method detection limit (MDL) here plotted equal to 0.61 µg/L, a lognormal distribution for the TCE concentration data might reasonably be assumed due to the linear trend above the MDL.

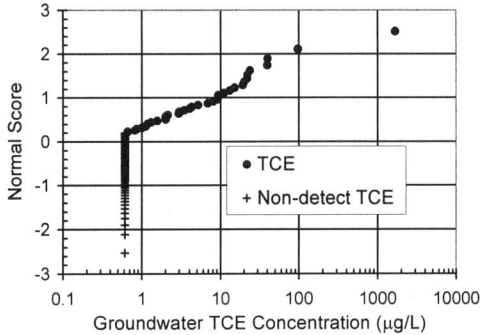

Figure 2. Normal probability plot of oil seepage basin TCE data.

To demonstrate the sample location selection method, a TCE contaminant plume that passes through each of the 73 Phase II values (using one-half the MDL for non-detect results) was generated using the sequential Gaussian simulation algorithm described by Deutsch and Journel (1992, page 141). The semivariogram estimated from the measured data as shown in Figure 3 was used. Logarithms of the measured TCE data were used as the basis for the simulation and the inverse logarithmic transform was applied to the sequential Gaussian simulated point values to get back to a lognormal distribution of TCE values. This simulated TCE surface is used as the true TCE plume from which samples are selected to illustrate our sample selection process and is not necessary to implement the sample selection method in practice.

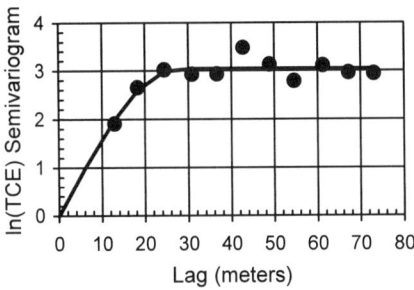

Figure 3. Semivariogram of the natural logarithm of the TCE (µg/L) concentrations.

Prior to use of the geostatistical based sample selection methodology, a semivariogram model needs to be developed. The initial semivariogram model and parameters might be based on prior experience, existing data, and/or physical characteristics of the problem at hand. The semivariogram should be periodically updated on the basis of additional data as it becomes available. In this case, sufficient data to estimate the semivariogram was not available at the outset of the investigation. Accordingly, a spherical semivariogram with a range of influence equal to 60 m, about one half the length of the source area (the trenches shown in Figure 1), was used. A value of 1 $[\ln(\mu g/L)]^2$, roughly twice the Phase I natural logarithm TCE variance, was used for the semivariogram sill. The intent here is not to justify this assumed variogram, but to suggest that initially, a variogram may need to be developed in the absence of adequate data. In practice, the variogram model should be updated as data become available.

Equation (1) was used to develop the probability map shown in Figure 4 on the basis of the 14 Phase I TCE values and a threshold T equal to the U.S. EPA Region III TCE Risk Based Concentration (RBC) for tap water of 1.6 µg/L. Regions with probabilities greater than 0.9 (p_2) or less than 0.2 (p_1) are shaded in Figure 4. Also shown shaded in Figure 4 are circles with a radius of 20 m (one third the assumed range of influence is used here to better define the variogram with subsequent data) centered on each of the existing 14 Phase I TCE sample locations. Unshaded regions in Figure 4 represent good candidate sample locations on the basis of our sampling strategy (between probability contours at $p_1 = 0.2$ and $p_2 = 0.9$ and greater than one third the range of influence, for this example, from existing data). Accordingly, 15 sample locations are suggested in this region, as shown in Figure 4. Note that the majority of these are hydraulically downgradient of the source area. To better define the source area, five samples were selected within the shaded area near the trenches (see Figure 4). Values for these 20 locations were selected from the simulated TCE plume. The probability map and variogram were updated using this new data and 20 additional sample locations were selected (Figure 5). Successive applications of the method were employed until 60 locations had been sampled resulting in the final probability plot shown in Figure 6. Successive sample locations may be selected one, or several, at a time.

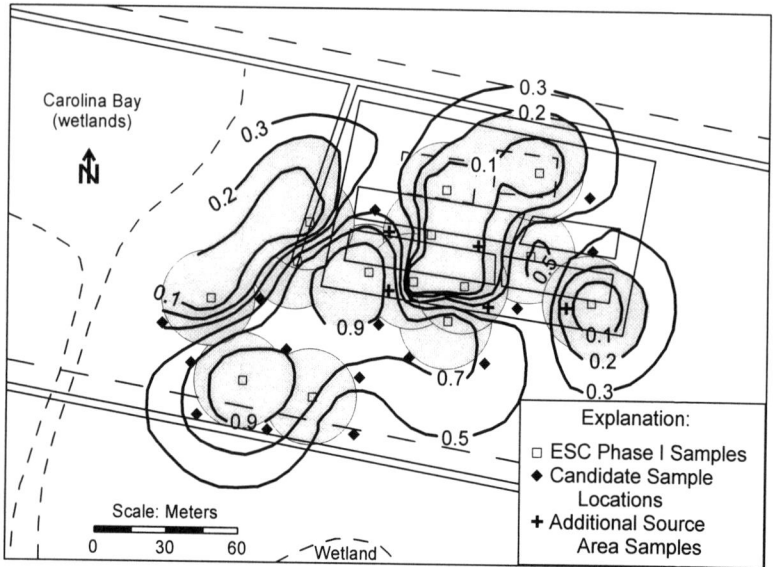

Figure 4.   Probability that TCE exceeds 1.6 ppb in groundwater based on the Phase I ESC data.

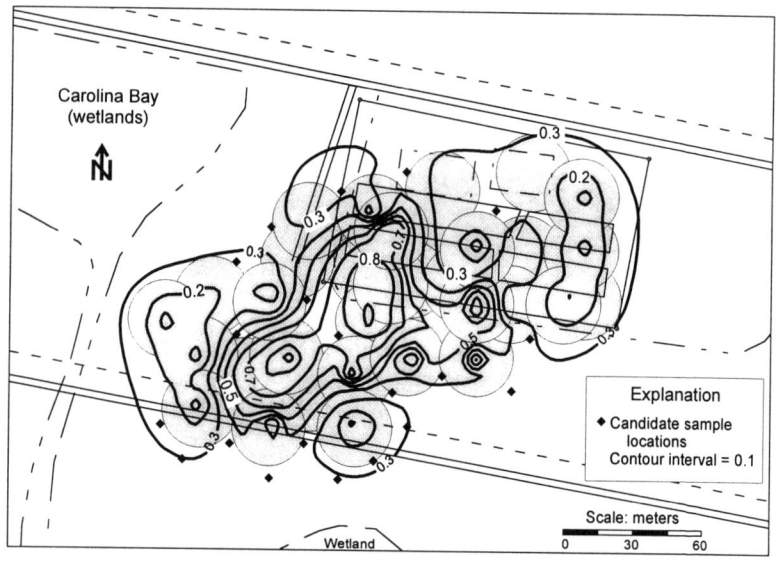

Figure 5.   Probability that TCE exceeds 1.6 ppb in groundwater after first sample event.

Because the detection limit (0.61 µg/L) is near the RBC (1.6 µg/L) and a portion of the data are non-detect, Stein's (1992) method was used to generate the final probability map shown in Figure 6. While Stein's method could be used at each sampling event, the method may require lengthy computer time (e.g. several hours) and ordinary or indicator kriging may be preferable for quicker field analyses. With the exception of the extreme southwest region of the plume (the wetland to the southwest of the site was a boundary beyond which further sampling was not allowed for this particular investigation), Figure 6 shows the TCE plume surrounded by a 0.1 probability contour indicating that the plume is well defined by these data at the level of the RBC threshold. Thus, further sampling is not necessary, according to our stopping rule, even though there are still large regions of the site with exceedance probabilities between 0.1 and 0.9.

Figure 7 shows the probability that TCE exceeds the RBC developed using Stein's method based on the 14 Phase I and 73 Phase II TCE data values collected during the ESC investigation. Comparing Figure 6 based on 14 Phase I and 60 TCE values selected by our method with Figure 7 based on 87 TCE values selected during the ESC investigation indicates that our method has reduced the uncertainty regarding the plume boundary locations using less data than was collected during the actual ESC investigation as a result of more effective placement of sample locations. While actual characteristics of the plume beyond the measured data are unknown, the plume boundaries were not well established with the ESC data. Application of geostatistical sample location methodology can be used to aid placement of samples to enhance the definition of a plume with a minimum of data. In practice, the methodology presented here could be applied to the simultaneous tracking of multiple contaminants by performing a similar analysis for each contaminant or by using indicator kriging by defining indicator values as $I(s_i) = 1$ if any individual contaminant exceeds its particular threshold at sample location s_i, and zero otherwise.

SUMMARY AND CONCLUSIONS

A geostatistical adaptive sampling method to define the spatial distribution of a contaminant plume whereby data are collected and analyzed in the field and used to determine subsequent sample locations within a flexible work plan that allows on-site decision making was developed. The method is intended to be used in conjunction with groundwater flow direction, local geology and other relevant information to aid sample location selection. The method uses site maps showing contours indicating the probability that a contaminant concentration exceeds a specified threshold, such as a regulated cleanup level, on the basis of the currently available data. Subsequent samples are selected from regions with greatest uncertainty regarding the likelihood that the contaminant exceeds the threshold. Sample spacing is about one third to one times the range of influence as defined by a semivariogram. Sampling continues until the contaminant plume is surrounded by a low probability contour, say 0.1, indicating those regions where contamination is not likely to exceed the threshold of interest.

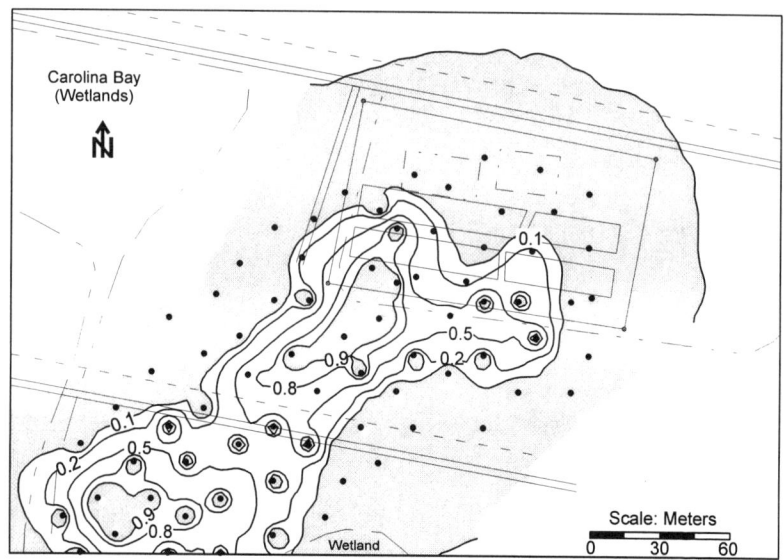

Figure 6. Probability that TCE exceeds 1.6 ppb based on our sample selection.

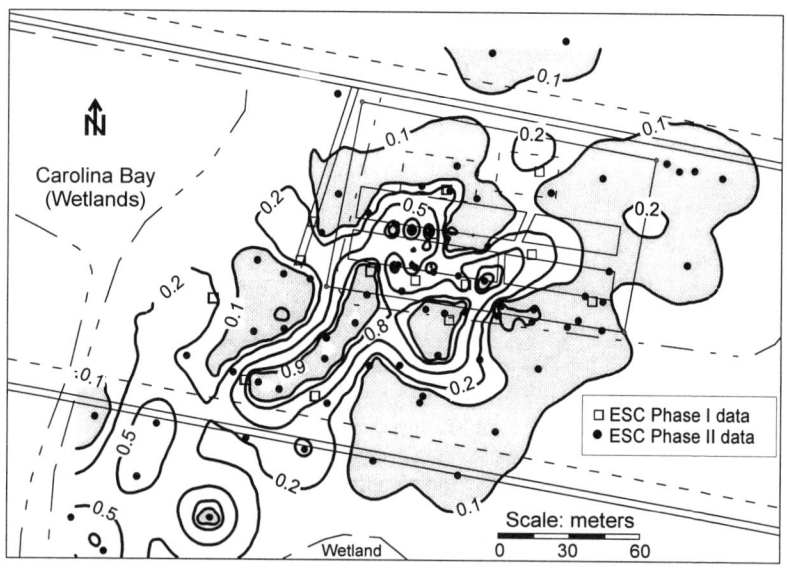

Figure 7. Probability that TCE exceeds 1.6 ppb based on the ESC samples.

If an assumed semivariogram is used to implement the process, then sampling should continue until a sufficient quantity and spacing of data are available to confidently estimate the semivariogram or covariance function. A data-based semivariogram, or covariance function, should be used to determine at least the final few sample locations and assess the validity of the geostatistical model. Sample spacing should be in accordance with characterization goals and the level of acceptable uncertainty because additional sampling to reduce uncertainty will increase characterization costs. Sample spacing at about one half of the range of influence should be adequate to implement this sampling strategy. Sample spacing at about the range of influence may be necessary to keep sampling and analysis costs down, however, some closer spaced sampling is required to properly estimate the semivariogram at low lag distances.

While the sample location selection and stopping rule described here were developed for an ESC program, whereby data are collected, analyzed, and interpreted in the field as the investigation proceeds, the geostatistical tools utilized have a broader application. These techniques can also be utilized in a more traditional site characterization to assess the level of uncertainty posed by existing data, plan the next phase of field sampling, and aid remedial decision making efforts by focusing attention on relevant uncertainties. Mathematical extension of the method to three spatial dimensions, rather than the two-dimensional plan view utilized here, is straightforward. We chose the two-dimensional approach because of its simplicity with respect to visualization of the three-dimensional problem. Cross sections and three dimensional computer generated images were also utilized in the field during this ESC project to examine the three-dimensional aspects of the plume as the site characterization progressed.

ACKNOWLEDGEMENTS

We are indebted to Ames Laboratory for providing us the opportunity to work with the Ames Laboratory ESC team at the Savannah River Site. Ames Laboratory is operated for the U.S. Department of Energy by Iowa State University under contract No. W-7405-Eng-82. This work was supported by the Office of Environmental Technology and Development.

REFERENCES

Burton, J. C., Walker, J. L., Jennings, T. V., Aggarwal, P., Hastings, B., Meyer, W. T., Rose, C. M., and Rosignolo, 1993, C. L., "Expedited Site Characterization: A Rapid Cost-Effective Process for Preremedial Site Characterization," Superfund XIV Conference and Exhibition, Greenbelt, Md., Nov. 30-Dec. 1, 809-826.

Cressie, N. A. C., 1991, *Statistics for Spatial Data*, John Wiley & Sons, Inc., New York.

David, H. T. and Yoo, S., 1993, Where Next? Adaptive Measurement Site Selection for Area Remediation, in *Environmental Statistics, Assessment, and Forecasting*, edited by C. R. Cothern and N. P. Ross, 361-372, Lewis Publishers, Inc., Michigan.

Deutsch, C. V. and Journel, A. G., 1992, *GSLIB: Geostatistical Software Library and User's Guide*, Oxford University Press, New York.

Englund, E. J. and Heravi, N., 1994, "Phased Sampling for Soil Remediation," Environmental and Ecological Statistics, 1, 247-263.

EPA, 1989, Methods for Evaluating the Attainment of Cleanup Standards, Volume 1: Soils and Solid Media, U.S. Environmental Protection Agency, EPA 230/02-89-042, February 1989.

EPA, 1992a, Statistical Methods for Evaluating the Attainment of Cleanup Standards, Volume 3: Reference-Based Standards for Soils and Solid Media, U.S. Environmental Protection Agency, EPA 230-R-94-004, December 1992.

EPA, 1992b, Methods for Evaluating the Attainment of Cleanup Standards, Volume 2: Ground Water, U.S. Environmental Protection Agency, EPA 230-R-92-14, July 1992.

EPA, 1993, Data Quality Objectives Process for Superfund: Interim Final Guidance, Office of Emergency and Remedial Response, U.S. Environmental Protection Agency, EPA/540/G-93/071, September 1993.

Isaaks, E. H. and Srivastava, R. M., 1989, *An Introduction to Applied Geostatistics*, Oxford University Press, Inc., New York.

Johnson, R. L., 1993, "A Bayesian Approach to Contaminant Plume Delineation," Proceedings of the Groundwater Modeling Conference, Golden, Colorado, June 9-12, E. Poeter, S. Ashlock, and J. Proud, eds., Colorado School of Mines, Golden, CO, 87-95, 1993.

Johnson, R. L., 1996, "A Bayesian/Geostatistical Approach to the Design of Adaptive Sampling Programs," *Geostatistics for Environmental and Geotechnical Applications, ASTM STP 1283*, R. Mohan Srivastava, Shahrokh Rouhani, Marc V. Cromer, Ivan Johnson, eds., American Society for Testing and Materials, Philadelphia.

Sandia National Laboratory, 1998, "Smart Sampling: Making sensible remediation decisions," Initiatives in Environmental Technology Investment, Waste Policy Institute, Volume 5, Fall, 1998.

Stein, M. L., 1992, "Prediction and Inference for Truncated Spatial Data," Journal of Computational and Graphical Statistics, Vol. 1, No. 1, 91-110.

Stenback, G. A., 1996, Application of Geostatistics to Expedited Site Characterization, Ph.D. thesis, 270 pp. Iowa State University, Ames, IA.

Stenback, G. A., Kjartanson, B. H., Bevolo, A., and Wonder, J. D., 1998, "Site Characterization, Expedited," *Encyclopedia of Environmental Analysis and Remediation*, edited by Robert A. Meyers, 4300-4329, John Wiley & Sons, Inc., New York.

Van Tooren, C. F. and Mosselman, M., 1997, "A Framework for Optimization of Soil Sampling Strategy and Soil Remediation Scenario Decisions using Moving Window Kriging," *geoENV I-Geostatistics for Environmental Applications*, edited by Amílcar Soares, Jaime Gómez-Hernandez, and Roland Froidevaux, Kluwar Academic Publishers, the Netherlands, 259-270.

GROUND IMPROVEMENT ASSESSMENT USING SCPTu AND CROSSHOLE DATA

By James A. Schneider[1] A.M. ASCE, Paul W. Mayne[2] M. ASCE,
and Glenn J. Rix[3] M. ASCE

ABSTRACT

A case study is presented where preloading was applied to reduce tank settlements at a liquid natural gas (LNG) facility in Puerto Rico. Stone columns were installed to further reduce settlements and provide an adequate factor of safety against soil liquefaction in the case of a seismic event. Seismic piezocone penetration tests (SCPTu) were used to quantitatively assess the effects of soil improvement. Soil properties estimated from SCPTu data taken before and after both improvement methods are compared to assess the degree of soil improvement. Preconsolidation stresses were estimated using methods based solely on net cone tip resistance, pore pressure difference, shear wave velocity, and a multiple regression based on cone tip resistance and shear wave velocity. Preloading provides additional vertical stress, which is directly related to an increase in preconsolidation stress if time rate consolidation properties are considered. Stone columns increase the lateral stress locally within the zone of influence of the column. Soils initially appear to be normally consolidated to slightly overconsolidated, and a significant increase in preconsolidation stress was noticed in clay layers after preloading. The high clay content of the soils likely led to a reduced zone of influence for the stone columns, and little improvement was noticed immediately after installation. Crosshole test (CHT) shear wave velocities taken across stone columns showed little to no increase in stiffness when compared to downhole shear wave velocities of the soil.

[1] Staff Engineer, GeoSyntec Consultants, 1100 Lake Hearn Drive, Suite 200, Atlanta, GA 30342
[2] Professor, Georgia Institute of Technology, School of Civil & Environmental Engineering, Atlanta, GA 30332-0355
[3] Director, Georgia Transportation Institute, and Associate Professor, Georgia Institute of Technology, School of Civil & Environmental Engineering, Atlanta, GA 30332-0355

INTRODUCTION

A liquid natural gas (LNG) terminal is being constructed in south central Puerto Rico, near Ponce. The main components of the terminal consist of two LNG storage tanks (east and west), a combustion turbine generator, a steam turbine, a heat recovery steam generator (HRSG), a HRSG stack, and a cooling tower. This case study is primarily concerned with the analysis of data from the East LNG tank.

Subsurface borings were initially performed to characterize soil conditions, layering, and variability. To gain more information about the initial soil properties and generate a more continuous soil profile, piezocone penetration tests (CPTu) were performed with pore pressure dissipation tests. Using information collected from standard penetration tests (SPT) and piezocone testing, soils in the subsurface profile were characterized as a mix of silty sand to clayey sand ($\approx$20-30% fines), with silty clay lenses. The soils in the upper 3.4 m below ground surface (BGS) are predominantly silty sands with clayey lenses. The soils between 3.4 and 9.1 m BGS were visually classified as sandy silts to lean clays. Estimation of soil behavior type from the Robertson et al. (1986) chart resulted in description of soils in the profile as silty clays to clays from 3.4 to 7.6 m and sensitive fines from 7.6 to 9.1 m (Figure 1). These soils show low tip resistance ($\approx$0.5-1.0 MPa), friction ratios of about 2-3 percent, and a pore pressure parameter (B_q) of about -0.25 to 2.0. At about 9.1 to 15.2 m below ground surface a cemented zone exists. The cemented zone is anticipated to act as an impermeable confining layer. The water table was initially at about 0.3 m, but 0.6 to 2.4 m of fill was added at the site prior to testing activities.

After an analysis of the initial soil conditions, two potential geotechnical problems were identified: (1) excessive settlements in the storage tank area; and (2) earthquake induced liquefaction of granular soils. This paper will be concerned with assessing the soil improvement methods with respect to excessive static settlements, not liquefaction. The soils between about 3.4 m and 9.1 m are of most concern in terms of settlements. To overcome these issues, a two-stage soil improvement program was developed. First the soils were preloaded to 125% of the design loads. After the surcharges were removed, stone columns were installed. Seismic piezocone penetration tests (SCPTu) were performed before the application of the surcharge, between the surcharging and the installation of the stone columns, and after installation of the stone columns. Crosshole shear wave velocity tests were also performed across the stone columns to assess their stiffness.

SOIL IMPROVEMENT PROGRAM

Two surcharge piles were placed at the site, one for each LNG tank. They were approximately 11 m high with a diameter of 91.4 m at the top of the pile. The additional vertical stress from the surcharge was approximately 200 kPa, no wick drains were installed to expedite consolidation, and the surcharge was left in place for about 11 months.

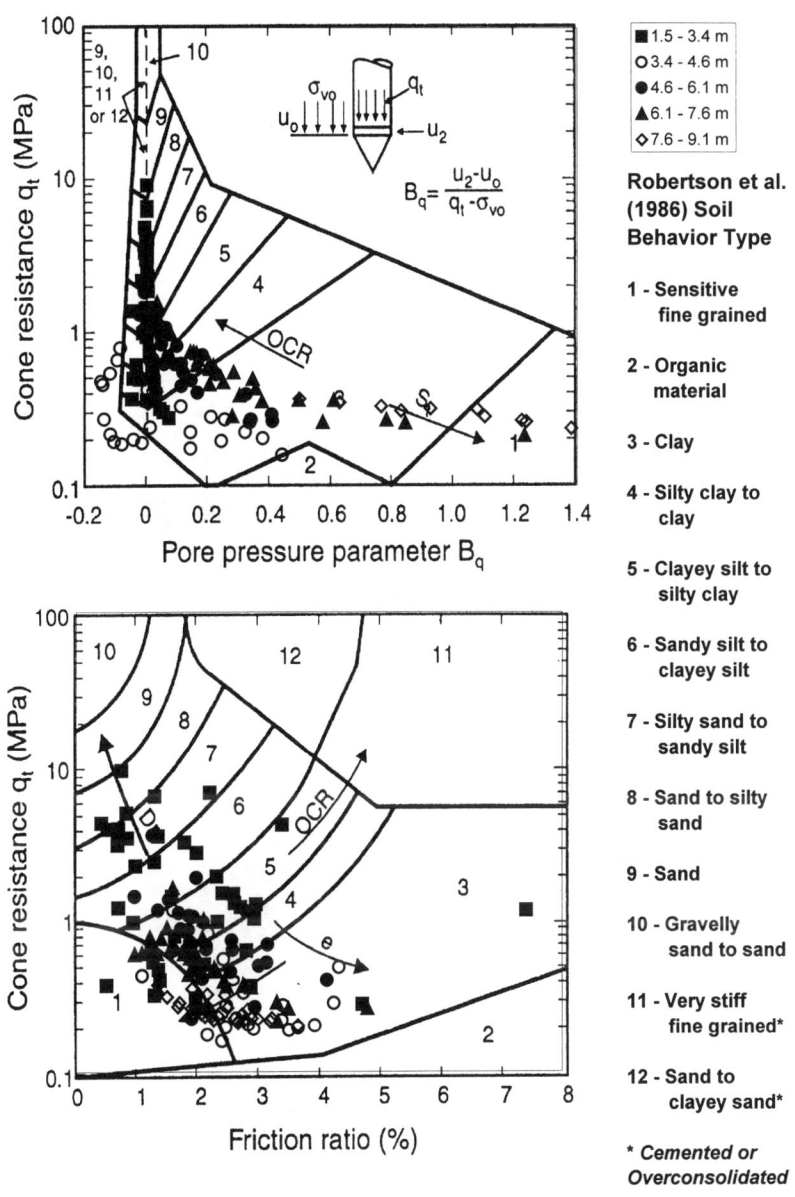

Figure 1. Pre-Construction CPTu Soil Behavior Type of Selected Layers

After the surcharge was removed, stone column installation began. There was a low
potential for improvement by densification since the soils contained 20 to 30 percent
fines, but the increase in stiffness of the soil - stone column matrix was anticipated to
provide sufficient reduction in tank settlements. This project used a 2.1-m on center
triangular spacing of 1.1-m diameter stone columns, resulting in a replacement ratio, A_r,
of 0.23. Each column was installed from the cemented layer at 9.1 to 15.2 m, and raised
to the surface using the dry bottom feed method. In-situ diameters were determined
using computer monitoring of amperages and stone quantity used.   Light towers
allowed installation 24-hours/day, resulting in completion of the stone column work in
six months. Figure 2 displays the site during the removal of a surcharge pile, and
Figure 3 shows testing at the site after installation of stone columns.

Figure 2. Site During Removal of Surcharge Pile

Figure 3. SCPTu Testing after Installation of Stone Columns

Soil improvement by surcharging will directly increase the preconsolidation stress (p'_c) by an increase in vertical effective stress. Soil improvement by stone columns should indirectly increase the preconsolidation stress. Additional lateral confinement from the vibro-replacement process is expected to increase the horizontal stress-state, and thus increase the lateral stress coefficient ($K_o = \sigma'_{ho} / \sigma'_{vo}$). This would in-turn increase the preconsolidation stress as related to the overconsolidation ratio ($OCR = p'_c / \sigma'_{vo}$) by the effective stress friction angle of the material (Mayne & Kulhawy, 1982):

$$OCR \approx \left(\frac{K_o}{1 - \sin\phi'} \right)^{1/\sin\phi'} \tag{1}$$

To quantitatively analyze soil improvement of clayey soils by surcharging and stone columns at this site, the preconsolidation stress is analyzed.

ESTIMATION OF PRECONSOLIDATION STRESS

The seismic piezocone provides four independent channels for analyzing effectiveness of soil improvement: (1) tip resistance, q_c; (2) sleeve friction, f_s; (3) pore pressure, u_2; and (4) shear wave velocity (V_s). Each of these measurements will depend on different soil properties to varying degrees, depending upon soil type and drainage conditions. Two derived parameters are also used in this study:

(1) Friction Ratio, $FR = f_s / q_t \cdot 100$; and
(2) Pore Pressure Parameter, $B_q = (u_2 - u_o) / (q_t - \sigma_{vo})$.

where u_o and σ_{vo} are equal to the in-situ pore pressure and total vertical stress, respectively, and the other parameters are as described above. To account for pore pressure effects acting on the back of the cone tip, the cone tip resistance (q_c) needs to be corrected to q_t using methods described in Lunne et al., (1997). Figure 4 shows seismic piezocone (SCPTu) data before and after preloading. Prior to SCPTu testing, the compacted fill was prebored and tests were started at about 1.5 m below the ground surface. Downhole shear wave velocity tests were performed at rod breaks, which corresponded to approximately 1.5 m intervals. Figure 5 displays SCPTu data before and after stone column installation. The bold line in Figure 4 refers to the same SCPTu as the bold line in Figure 5. Crosshole shear wave velocity values presented in Figure 5 were collected by performing tests across the stone columns, and will be discussed later.

Seismic piezocone data were analyzed using four correlations that relate SCPTu parameters to the preconsolidation stress (p'_c). Methods based solely on net tip resistance and pore pressure difference were used (Chen & Mayne, 1996), as well as methods based on shear wave velocity and a multiple regression including net tip resistance and shear wave velocity (Mayne et al., 1998). While there is scatter inherent in these correlations from variations in soil deposits, the relative increase in p'_c at this site should be appropriate to assess the degree of soil improvement. Calibration of the

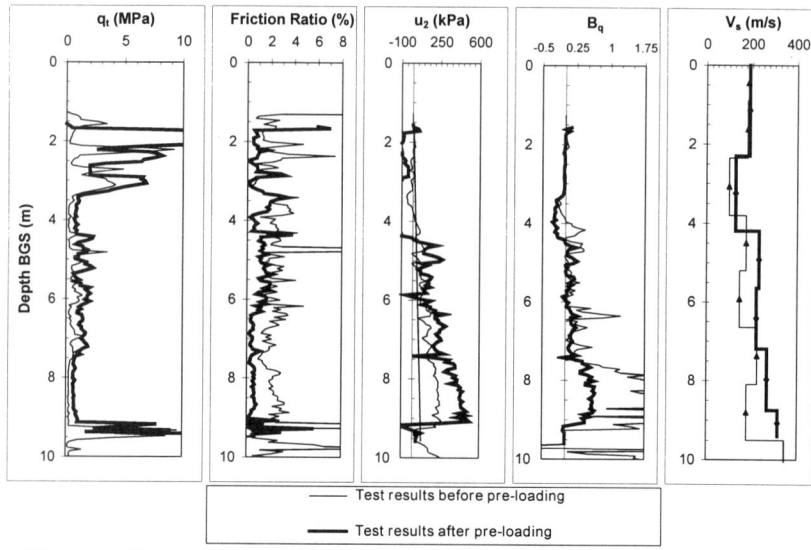

Figure 4. Comparison of SCPTu Data Before and After Preloading at LSP-16

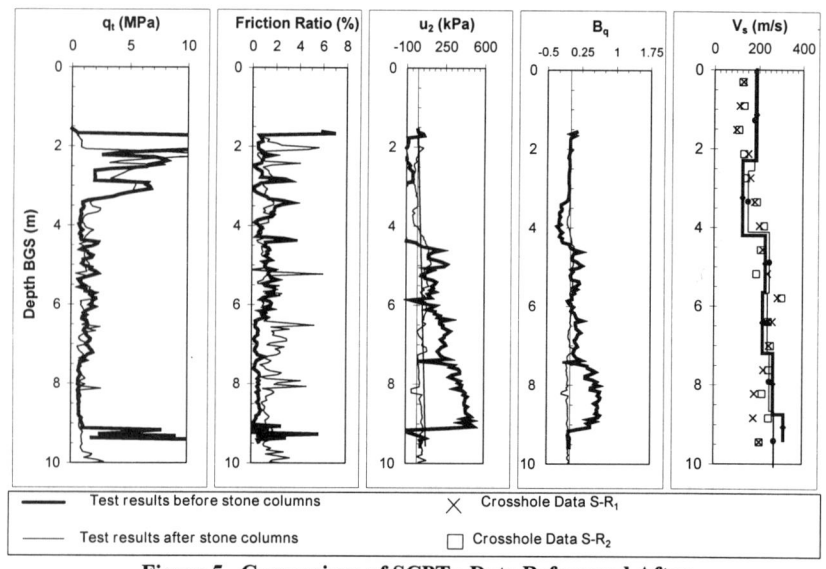

Figure 5. Comparison of SCPTu Data Before and After Stone Column Installation at LSP-16

correlations with laboratory data is recommended if settlement calculations are preformed using p'_c from the following correlations:

$$p'_c = 0.31 \cdot (q_t - \sigma_{vo}) \tag{2}$$

$$p'_c = 0.53 \cdot (u_2 - u_o) \tag{3}$$

$$p'_c = (V_s/4.59)^{1.47} \tag{4}$$

$$p'_c = (q_t - \sigma_{vo})^{0.702} \cdot (V_s/64)^{0.751} \tag{5}$$

where stress terms are in kPa, and V_s is is m/s. These four correlations based on three independent parameters were used for the analyses, as shown in Figure 6. Good agreement was obtained with all methods except the pore pressure difference. It is expected that the silt and sand content led to partially drained conditions, thus reducing the pore pressure response (Lunne et al., 1997) and effectiveness of that correlation (Eq.

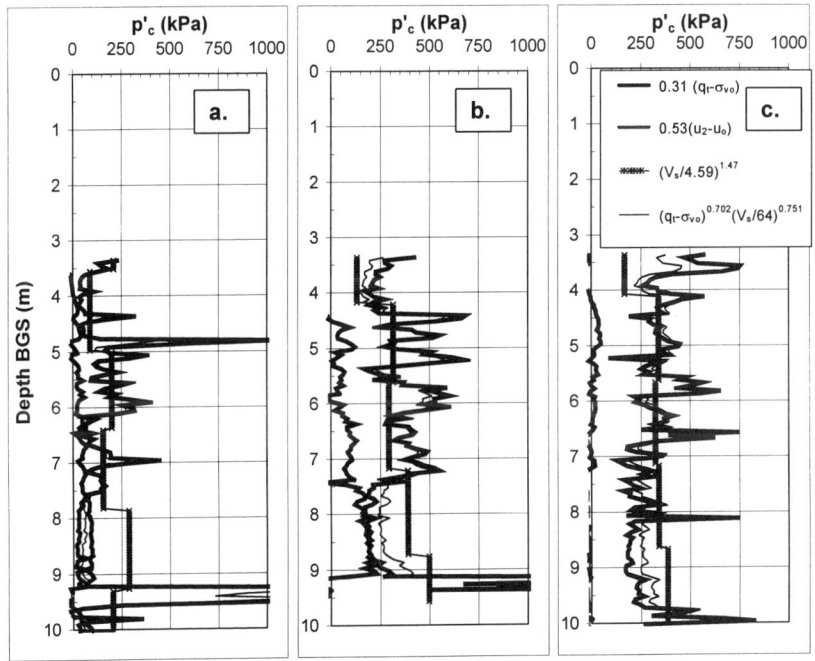

Figure 6. Estimations of Preconsolidation Stress
(a) Before Improvement (b) After Preload (c) After Stone Columns

3). It should be noted that according to Robertson et al. (1986), as OCR increases the tip resistance should increase and the pore pressure parameter should decrease. While the pore pressure did not behave in a pure undrained condition, the pore pressure parameter did behave as expected when the preconsolidation stress increased between tests. Table 1 displays the preconsolidation stress selected from the trends shown in Figure 6. The selected values were based on a representative average of the four correlations. If B_q was less than 0.4 (Senneset et al., 1988), the penetration was considered partially drained and the pore pressure difference correlation (Eq. 3) was not considered.

Table 1. Estimated Preconsolidation Stress from SCPTu
Before and After Soil Improvement

Depth (m)	Before Improvement (kPa)	After Preloading (kPa)	After Stone Columns (kPa)
0 - 3.4 silty sand with clay lenses	NA	NA	NA
3.4 - 4.9 clay	90	245	335
4.9 - 6.4 silty clay	200	380	360
6.4 - 7.9 clay	140	350	320
7.9 - 9.4 sensitive fines	85	130	170

NA - Not Applicable to Analyses

STIFFNESS OF SOIL COMPARED TO STONE COLUMNS

Since these soils contained relatively high fines contents (> 20%), an increase in relative density from the vibro-replacement stone columns would not be apparent. The stone columns primary use is as a stiffener to reduce settlements. Therefore, a measure of the stone column shear modulus as compared to the soil shear modulus was desired. In downhole seismic testing (SCPTu), the shear wave is horizontally polarized and

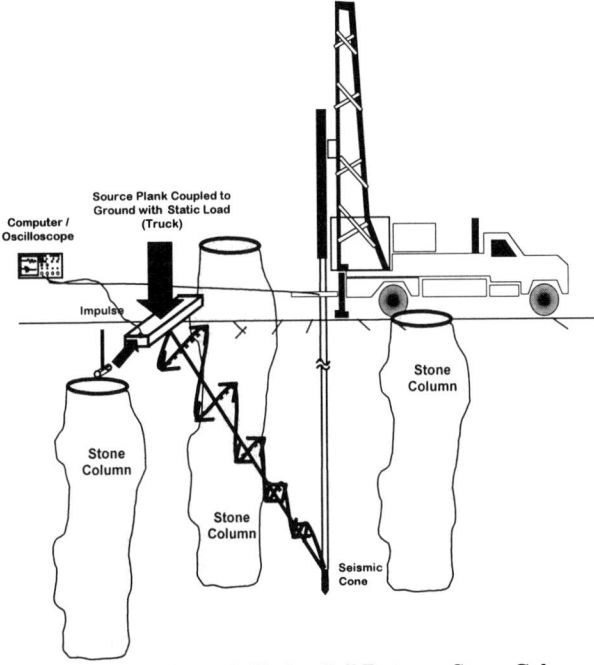

Figure 7. SCPTu through Native Soil Between Stone Columns

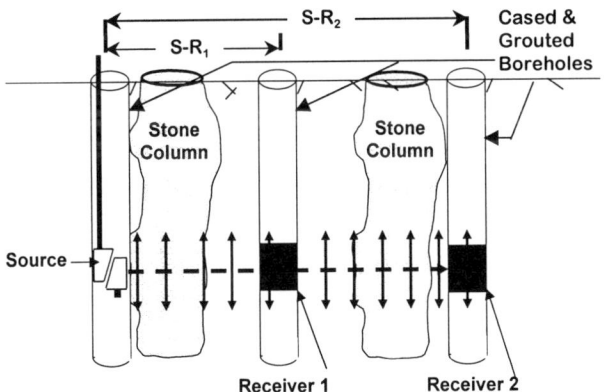

Figure 8. Crosshole testing through Stone Columns

vertically propagating. This arrangement is good for horizontally stratified soils, but not the vertical arrangement of stone columns. When V_s measurements of native soils are attempted between stone columns, the chance exists that stone column material may influence the downhole seismic wave (Figure 7). In crosshole seismic testing, the wave is typically vertically polarized and horizontally propagating. This will produce a wave with vertical particle motion as it passes through a stone column, the desired configuration. The selected borehole arrangement is used to measure the stiffness of two stone columns with one impulse to check repeatability (Figure 8). Other arrangements can penetrate the treated soil and the stone column with one impulse, to assess variability of stiffness between the two mediums (ASCE, 1997). The close spacing of the stone columns at this site did not accommodate a borehole arrangement that tested the native soil and stone column with one impulse.

Once the shear wave velocity is known, it can be used directly to compute the soil or stone column shear modulus or small strain stiffness ($G_{max} = \rho V_s^2$). There has been much work relating shear wave velocity to void ratio and in-situ density, thus the small strain stiffness can be related directly to shear wave velocity. A statistical relationship has been presented by Mayne et al. (1999) relating total mass density data from gravels, sands, silts, and clays, to shear wave velocity, V_s, in m/s and depth, z, in meters (n = 727; $r^2 = 0.730$):

$$\rho_t \approx 1 + \frac{1}{0.614 + 58.7(\log z + 1.095)/V_s} \tag{6}$$

Figure 5 displays the shear wave velocity of the soil (from SCPTu) and the shear wave velocity of the stone column (from crosshole tests). The shear wave velocity and stiffness of the stone column and soil are not significantly different. This will lead to a stiffness ratio, G_r, of about 1.

It should be noted that the crosshole tests were performed within about 1 month of the installation of the stone columns. Disturbance to the clay structure during installation of the stone columns may reduce the lateral confinement, and thus reduce the shear wave velocity through the column. While crosshole shear wave velocity was generally slightly higher than SCPTu shear wave velocity in the silty clays between 3.4 and 7.6 m, the crosshole shear wave velocity was slightly lower than SCPTu shear wave velocity in the sensitive fines between 7.6 and 9.1 m. The reduction in lateral confinement in the sensitive fines may have been greater than that in the silty clays.

CONCLUSIONS AND RECOMMENDATIONS

The initial soil improvement phases of this project have been completed, but construction of the tanks and supporting structures is currently ongoing. Quantitative analysis of preloading for soil improvement shows an increase in preconsolidation stress by a factor of 1.9 to 2.7. Assessment of additional improvement from installation of stone columns resulted in a preconsolidation stress that was 0.9 to 1.4 times the post preloading preconsolidation stress. As expected, for the silty clay soils analyzed at this site preloading provided a more substantial increase in estimated preconsolidation stress than the additional lateral confinement associated with stone column installation. At depths between 4.6 and 7.6 m, it appears that the preconsolidation stress may have decreased by up to 10 percent. This apparent reduction in p'_c was likely a result of either local variability, or a reduction in SCPTu parameters due to local disturbance induced by the stone column installation. The primary function of the stone columns was to provide an increase in matrix stiffness, as well as additional factor of safety against soil liquefaction in the case of an earthquake event. Shear wave velocity tests were compared to assess relative stiffness of the soil and stone columns.

The shear wave velocity measurements taken during this study resulted in approximately equal stiffness values for the soil and stone columns. This was unexpected, and a number of test related issues call for additional testing:

1. The downhole readings may have been influenced by the shear wave passing through the stone column (Figure 7).
2. The crosshole wave was influenced by the shear wave passing through the native soils and coupling between the soil and stone column (Figure 8).
3. Pore pressures induced in the native clays from stone column installation may have reduced the lateral confinement of the stone columns, and thus may have resulted in a reduced the shear wave velocity. Time between the end of column installation and beginning of crosshole testing may not have been sufficient for dissipation of the pore pressures in the confining clay.

These issues indicate that testing at a later date and in a different configuration may provide more representative data concerning the stiffness ratio of the soil and stone columns. Gaps between the stone column and boreholes, and coupling issues between the soil and stone columns may result in the stiffness of the soil influencing the interpreted stiffness of the stone column. The presence of gaps between the stone column and borehole are inevitable during testing, due to the unknown variability in diameter of the column with depth. The setup where the crosshole shear wave is generated in a first borehole, passes through native soil to a receiver in a second borehole, and then passes through the stone column to a receiver in a third borehole (ASCE, 1997) may be the most appropriate configuration to assess relative stiffness of the soil and stone column. Depending upon the spacing and configuration of the stone columns, this arrangement may not be a feasible option, as was the case at this site.

ACKNOWLEDGMENTS

The authors acknowledge the support of the NSF geotechnical program for which Dr. Priscilla P. Nelson and Dr. Mehmet T. Tumay were the successive program directors. The project team worked with GeoCim, Enron, Black & Veatch, and Dames & Moore during the improvement phases discussed herein.

REFERENCES

ASCE (1997) Vibro-Stone Columns, *Ground Improvement, Ground Reinforcement, Ground Treatment: Developments 1987-1997, Chapter 2.1*, GSP 69, ASCE, Reston, Virginia, pp. 74-91.

Chen, B.S-Y., and Mayne, P.W. (1996) Statistical relationships between piezocone measurements and stress history of clays, *Canadian Geotechincal Journal*, 33 (3), pp. 488-498.

Lunne, T., Robertson, P.K., and Powell, J.J.M. (1997). *Cone Penetration Testing in Geotechnical Practice*, Blackie Academic & Professional, New York; available from E&FN SPON/Routledge, New York.

Mayne, P.W., and Kulhawy, F.H. (1982). K_o – OCR relationships in soil, *Journal of Geotechnical Engineering Division*, 108 (GT6), pp. 851 – 872.

Mayne, P.W., Schneider, J.A., and Martin, G.K. (1999). Small- and large-strain soil properties from seismic flat dilatometer tests, *Pre-failure Deformation Characteristics of GeoMaterials, Vol. 1*, Balkema, Rotterdam, pp. 419 – 426.

Mayne, P.W., Robertson, P.K., and Lunne, T. (1998). Clay stress history evaluated from seismic piezocone tests, *Geotechnical Site Characterization, Vol. 2*, Balkema, Rotterdam, pp. 1113-1118.

Mitchell, J.K. (1986). Ground Improvement Evaluation by In-Situ Tests, *Use of In-Situ Tests in Geotechnical Engineering*, GSP 6, ASCE, Reston, Virginia, pp. 221-236.

Robertson, P.K., Campanella, R.G., Gillespie, D., and Greig, J. (1986). Use of piezometer cone data, *The Use of In-Situ Tests in Geotechnical Engineering*, GSP 6, ASCE, Reston, VA, pp. 1263-1280.

Senneset, K., Randven, R., Lunne, T., By, T., and Amundsen, T. (1988). Piezocone tests in silty soils, *Penetration Testing 1988*, Vol. 2, Balkema, Rotterdam, pp. 955-966.

COMBINATIONS OF IN-SITU TESTS FOR CONTROL OF GROUND MODIFICATION IN SILTS AND SANDS

By John A. Howie[1], Chris Daniel[2], Ali Amini Asalemi[2], and R.G. Campanella[3],

ABSTRACT

This paper investigates the use of in situ test data to characterize the changes in stress and density induced by ground improvement. Testing comprised seismic cone penetration tests, full displacement pressuremeter tests and resistivity cone penetration tests. After ground treatment, changes were observed in tip resistance, pore pressure response, shear wave velocity, the characteristics of pressuremeter curves and bulk resistivity. Some of these changes can be caused by changes in lateral stress as well as by density increases. The results generally indicate that the use of a combination of in situ tests will improve our understanding of the changes in soil behaviour achieved by ground treatment. The improved understanding of soil behaviour obtained from combinations of tests may allow ground improvement specifications to be written in terms of a desired post-treatment stress-deformation response. However, further research is required to develop reliable methods of characterization of stress-deformation behaviour.

INTRODUCTION

Vibratory compaction of granular deposits and reinforcement of silts and clays by stone columns are methods of ground improvement used to reduce the potential for earthquake-induced ground movements in the soils of the Fraser River Delta around Vancouver, British Columbia. These ground movements will be the result of softening due to pore pressure generation and redistribution, and liquefaction. In sands, in situ testing by Standard Penetration Test (SPT) or Cone Penetration Test (CPT) is commonly used to assess the need for ground improvement and to check the degree of improvement obtained. In recent years, shear wave velocity (V_s) has also been used for

---

[1] Associate Professor (Email: jahowie@civil.ubc.ca),
[2] Graduate Student, Geotechnical Engineering, and
[3] Professor Emeritus, Department of Civil Engineering, University of British Columbia, Vancouver, B.C., V6T 1Z4 Canada (Email: rgc@civil.ubc.ca).

this purpose (NCEER, 1997). In liquefiable silts, such performance criteria are difficult to apply, as design methods require large corrections to measured penetration resistances (Robertson and Wride, 1997). In soft clays, such methods are not applicable and it is necessary to use other techniques such as the "Chinese Criterion" (Wang, 1979) to assess liquefaction susceptibility.

This paper presents the results of seismic piezocone (SCPTU), full-displacement pressuremeter (FDPM) testing and resistivity piezocone (RCPTU) testing carried out in potentially liquefiable sands and silts improved by vibro-replacement (stone columns). These results are used to illustrate important issues in the use of penetration testing for quality control of ground improvement and to examine the potential benefits of the use of combinations of in situ tests for this purpose.

GROUND IMPROVEMENT BY VIBRO REPLACEMENT

Cohesionless soils are densified most effectively by vibratory means. In vibro-replacement, a vibrating rod is inserted into the ground and coupling between the vibrator and the surrounding ground results in the soil being vibrated into a denser packing. At regular depth intervals, stone is added until no further improvement in resistance to vibration is observed. The use of stone allows probes to be placed at larger spacings than if densification was by vibration alone. The stone columns provide an additional benefit by acting as zones of higher permeability for dissipation of any pore pressures created by seismic shaking. The degree of densification achieved depends on the level of energy applied and the pattern and sequencing of its application.

As the proportion and plasticity of fines in the soil increase, the effectiveness of vibratory compaction becomes lower. Increasing fines content reduces the permeability of the soil being treated. This restricts drainage and makes it more difficult to force the soil particles into a denser packing. When high pore pressures are generated or liquefaction occurs in the soil adjacent to the soil being vibrated, the transmission of additional vibration to the surrounding soil is severely attenuated. The radius of influence of the process is thus greatly reduced and treatment procedures must be adjusted. The spacing of the treatment locations must be reduced and the treatment must be sequenced to allow time for dissipation of excess pore pressures before further compactive effort is applied. Other forms of improvement may be more appropriate. In these soils, improvement of the resistance of the soil to seismic shaking can be achieved by considering the reinforcing action provided by the presence of stone columns.

QUALITY CONTROL OF GROUND IMPROVEMENT

Specifications for ground improvement are typically written in terms of the achievement of a required performance. For seismic design, ground treatment is conventionally carried out to meet a performance criterion based on penetration resistance, typically selected based on the chart shown in Figure 1 (NCEER, 1997). This relates the CPT tip resistance normalized to a standard stress level of 100 kPa, $q_{c1N,}$ to the Cyclic Stress Ratio (CSR) to cause liquefaction, using the expression:

$$q_{c1N} = (q_c/p_a)(p_a/\sigma_v')^{0.5} \tag{1}$$

where σ_v' is the vertical effective overburden stress and p_a is 100 kPa or approximately one atmosphere. The CPT tip resistance, q_c, should strictly be corrected to q_t to account for pore pressure effects on unequal end areas, but this correction is only important in silts and clays where tip resistances are low relative to the pore pressures around the tip (Campanella et al., 1982).

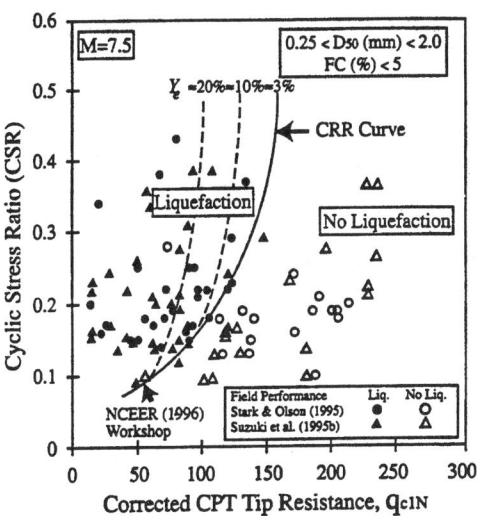

Figure 1 - Curve Recommended for Calculation of CRR from CPT Data along With Empirical Liquefaction Data (After Robertson and Wride, 1997)

The ground improvement is typically required to achieve a penetration resistance that will fall well to the right of the Cyclic Resistance Ratio (CRR) curve. Mitchell et al. (1998) notes that sites improved to such performance criteria have performed well in recent earthquakes. For non-seismic design, the required performance may be a maximum allowable settlement at working load and this may be assessed by the performance of a load test at the completion of the contract.

The presence of increased fines content affects the assessment of the degree of improvement achieved. In silts and clays, the tip resistance is a relatively insensitive measure of improvement in strength and stiffness. Campanella et al. (1982) interpreted changes in pore pressure response during penetration from positive to strongly negative to indicate improvement in the liquefaction resistance of silts. During site exploration, it is frequently difficult to distinguish between layers of loose sands and soils with high silt contents.

Pressuremeter (PM) testing has also been used to assess ground improvement. Indeed, Debats et al. (1998) argue that use of the pressuremeter in accordance with the Menard rules, is the optimum method for assessment of ground improvement, particularly in soils with fines. They argue that the results are more sensitive to stress history than are the results of CPTU testing and shear wave velocity measurements. However, for improvement against liquefaction, there is little guidance as to the relationship between the results of the static pressuremeter test and the response of the soil to cyclic loading. Cyclic pressuremeter testing may offer some insight but this development is still in the research phase.

FIELD TESTING

Soil Conditions and Performance Specification

Field-testing before and after ground improvement by vibro-replacement was carried out at two sites. Soil conditions in the Delta typically consist of fine-grained overbank, floodplain and intertidal clayey silts overlying Fraser River sands, which in turn overlie marine foreslope deposits, consisting of loose to compact silty sands interbedded with soft to firm grey clayey silt (Armstrong, 1981). Fraser River Sand is a fine to medium uniform sand. Based on its depositional history, it can be expected to contain zones with variable amounts of non-plastic fines and lenses of silt.

At the first site, ground improvement was done after site preparation by stripping of a thin layer of organic material and replacement with a sand-fill working platform. The surficial clayey silt was not removed. At the second site, most of the surficial silt was removed prior to ground improvement. At each site, the contractor was required to select the equipment, stone column spacing and construction procedure to achieve the required performance. The performance was assessed by CPTU testing at points equidistant from the columns, i.e. at the centroid of the triangles of stone columns. The consultant selected test locations based on field review of construction.

In Situ Testing

The fieldwork reported in this paper was carried out separately from the contract *in situ* testing. At the first site, the objective was to investigate whether a Seismic Cone Pressuremeter (SCPM) would provide additional information for quality control of vibro-replacement over the more standard CPTU. The ground improvement work was already under way when the research was initiated and so the "before" data were obtained in an area just off site which had not experienced the site preparation undertaken on the site proper. The "after" testing was carried out about 15 metres away in a zone where vibro-replacement had been carried out to 10 m below original grade. The testing at each location consisted of a seismic piezo-cone sounding (SCPTU) using a standard 35.7 mm cone followed by a 74 mm diameter full-displacement pressuremeter (FDPM) test in the same hole. The FDPM testing was carried out by pushing the University of British Columbia (UBC) Self-Boring Pressuremeter (SBPM) behind a conical tip of the same diameter as the SBPM. The pressuremeter expansion test is thus not strictly comparable to a cone pressuremeter test as considerable time elapsed between the SCPT sounding and the pressuremeter test. However, as the same procedure was used before and after improvement, the results should be comparable. The equipment and procedures followed for the seismic CPT testing are as described in Robertson et al. (1986). Hammer blows on the ends of a beam supporting the truck were used to generate shear waves. The expansion section of the UBC SBPM has a length to diameter ratio of about 6. It is a monocell probe with displacement of the membrane measured using six strain arms spaced equally around the circumference of the probe at mid-height of the expanding section. The tests were carried out by inflating the membrane at a constant rate to 3% strain. The pressure was held constant for 5 minutes, and then unload-reload loops were carried out before the pressure was again increased. In some cases, additional unload-reload loops were carried out at higher strain levels prior to unloading.

The objective at the second site was to investigate the use of a combination of shear wave velocity and penetration resistance for assessment of ground improvement. Based on elastic theory, the small strain shear modulus G_o, can be determined from the shear wave velocity, V_s, using the expression:

$$G_o = \rho V_s^2 \tag{2}$$

where ρ is the soil mass density. In addition, a resistivity module was included behind the SCPTU. The UBC resistivity module consists of two pairs of brass ring electrodes spaced at 150 and 15 mm within an insulating plastic (Daniel et al., 1999). The applied potential difference to establish a 1000 Hz excitation current of 0.025, 0.25 or 2.5 mA across the outer electrodes is recorded. Excitation magnitude is operator-controlled during the sounding. The recorded voltage is correlated to bulk soil resistivity using calibration factors obtained in fluids of known resistivity.

At the second site the soundings were carried out in almost exactly the same location before and after densification to limit the effects of soil variability on the results. It was considered that the ground improvement would largely erase the effects of the previous CPT sounding due to overlap of the zones of densification. This was qualitatively confirmed by the testing. The tip resistance increased greatly in the improved zone but the "after" tip resistances were considerably lower than the "before" data below the improved zone. Care was taken to avoid placement of the pad used as the shear wave source over any stone columns to minimize the potential for shear waves travelling preferentially through the stone column.

RESULTS

Effect of Ground Improvement on Cone Penetration Test Results

Figure 2 shows the CPTU profiles measured before and after ground improvement at the first site. The "after" tests were carried out about 3 months after vibro-replacement. The effect of the ground treatment in the sand below about 6 m is readily apparent with

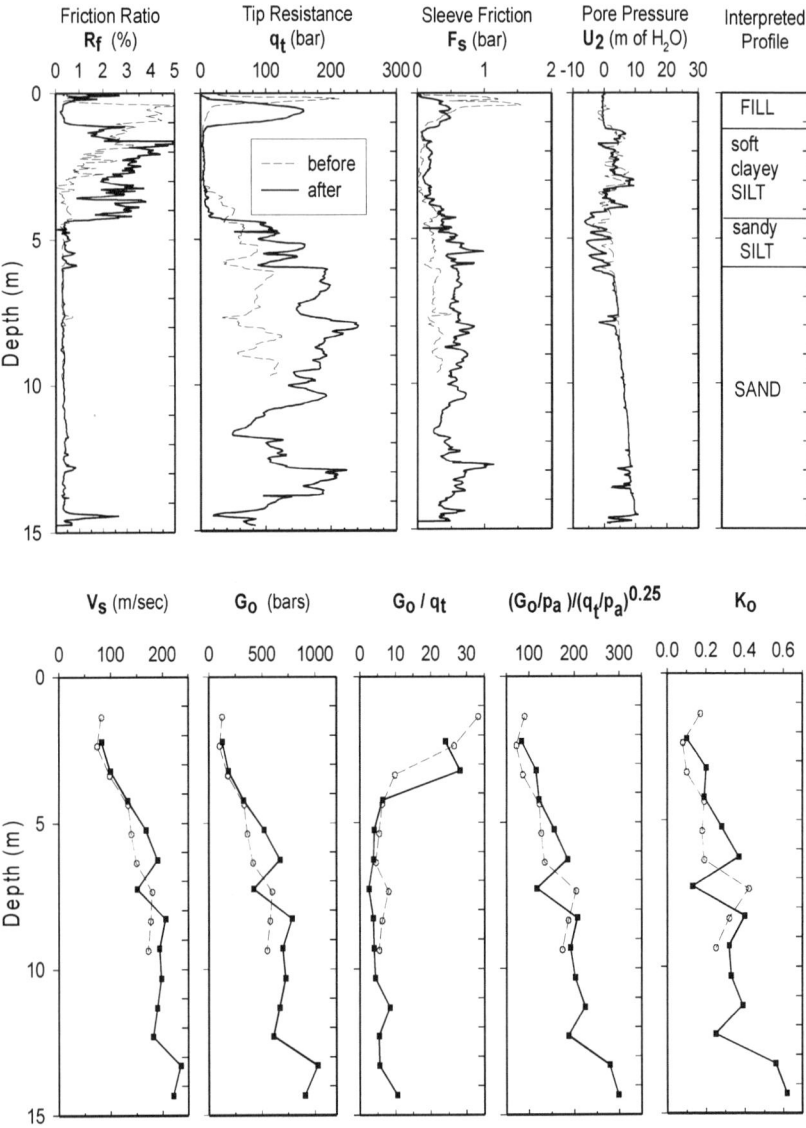

Figure 2 – SCPTU Profiles Before and After Vibro-Replacement – First Site

the tip resistance having increased by up to 117%. The friction value has also increased and the pore pressure response in the upper sand has become more strongly negative, i.e. below equilibrium, indicating a greater tendency for the sand to dilate. The negative pore pressure response could be interpreted to mean that there is elevated silt content in the zone between about 4.5 m and 6 m below ground but the level of improvement achieved suggests that the soil is reasonably clean.

The level of improvement in the clayey silt from 1 to 3 m, indicated by comparison of q_t values before and after treatment, is inconclusive but there appears to have been an increase in friction values. Figure 3 shows the cone profile at an enlarged scale to allow comparison of the effect of the ground treatment on the soft silt. There has been an increase in the tip resistance and an apparent doubling of sleeve friction suggesting that the treatment has been effective in increasing the strength of the silt but the difference in readings is close to the resolution of the instrument. The pore pressure response before treatment indicated rapid variation in the dilation characteristics of the soil, likely a result of interbedding of silts, sands and silty sands. The post-treatment response is more consistently positive. This may be a consequence of site variability but could also be the result of mixing of the layers by ground treatment.

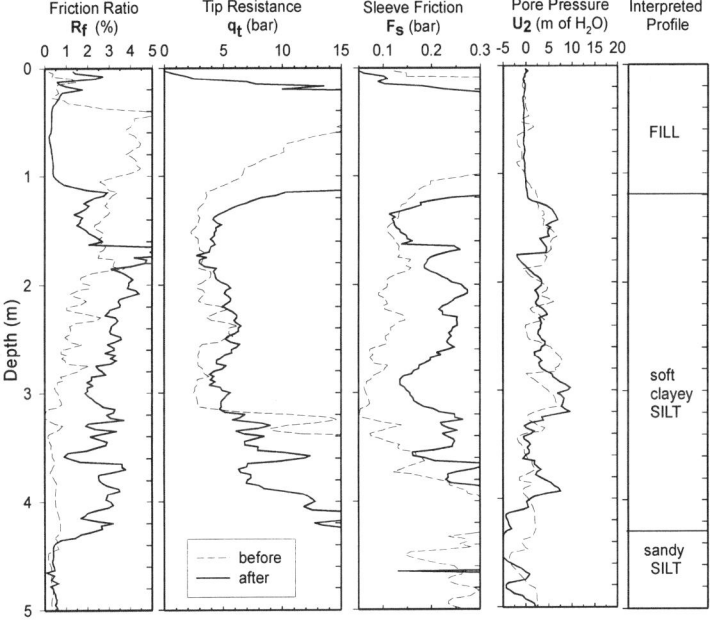

Figure 3 – CPTU Profiles for first site at expanded scale

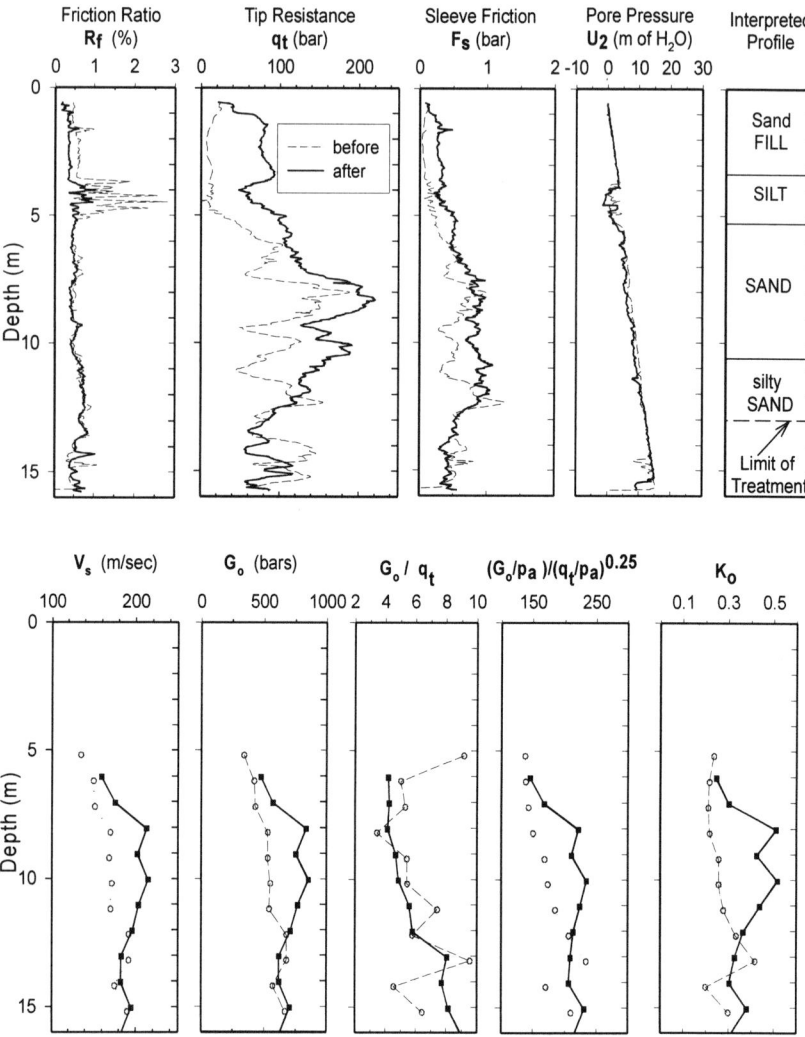

Figure 4 – SCPTU Profiles Before and After Vibro-Replacement – Second Site

Figure 4 from the second site, shows a similar large increase in tip resistance achieved by the ground improvement down to 13 metres below ground, the maximum depth of

treatment. The increase is most noticeable in the loose sand fill and underlying sand above about the three metre depth. In this layer, the friction ratio was reduced as a result of densification although the material did not change. Also of interest is the large increase in tip resistance achieved in the material between 3.5 m and 5 m depth. This material had layers with friction ratios over 2% before densification. The friction ratios observed after densification were below 1%.

Effect of Ground Improvement on Shear Wave Velocity

Figures 2 and 4 also show the values of V_s measured before and after vibro-replacement. At both sites, there is a clear increase in V_s in the sands but the increase in the clayey silts at the first site is less. This may be because clayey silts require a much longer time interval than the sands to recover from the effects of the remoulding experienced during ground treatment. The low value at 7 m depth in Figure 2 is likely due to a localized silt zone.

Effect of Ground Improvement on Pressuremeter (PM) Curves

Figures 5 and 6 present the results of PM tests before and after ground treatment in the sands and clayey silts, respectively. The capacity of the pushing equipment was

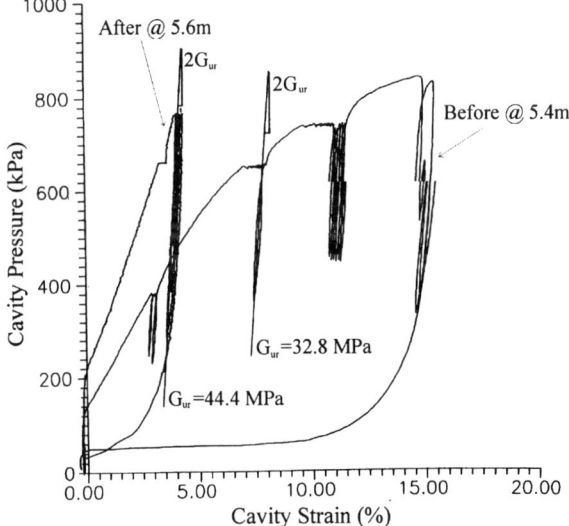

Figure 5 – Pre- and Post-Densification Pressuremeter Tests in Sand

insufficient to push the pressuremeter unit deeper than 5.5 m (about 1.2 m into the densified sand). At this depth, it was not possible to inflate the pressuremeter past a cavity strain (change in radius/initial radius) of about 3% due to the limitations of the pressure system.

In the sand, the measured lift-off pressure increased from about 140 to 220 kPa - an increase in effective lateral stress from about 90 to 170 kPa or an increase of about 90%. The slope of the initial part of the expansion curve during the initial phase of expansion was much steeper after improvement (7 MPa versus 4 MPa) and the slope of the unload-reload loops increased from an equivalent shear modulus of about 33 MPa to 44 MPa.

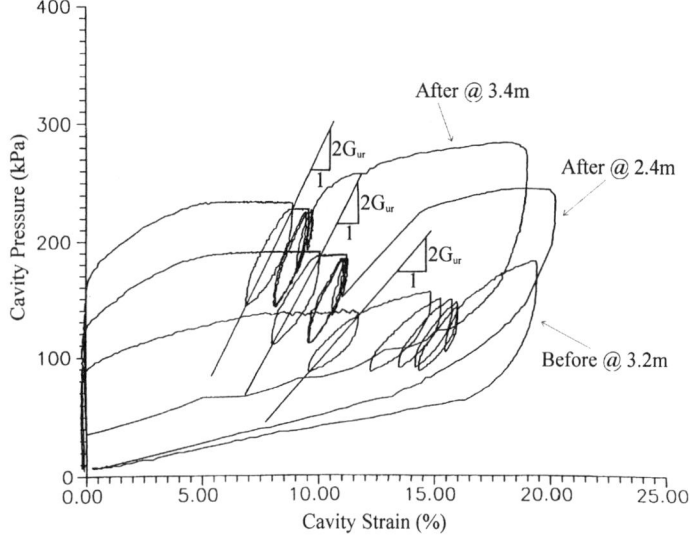

Figure 6 – Pre- and Post-Densification Pressuremeter Tests in Clayey Silts

In the silts, the lift-off pressures increased. There was no pore pressure measurement adjacent to the membrane so the proportion of the lift-off pressure made up of pore pressure is unknown. There has also been an increase in the maximum pressure attained and in the distance between the expansion curve and the contraction curve. This distance varies with shear strength in SBPM tests (Wroth, 1984) and is assumed to be an indicator of shear strength in Figure 6. It appears that there has been an increase in shear strength due to the vibro-replacement. The unload-reload stiffnesses have also increased from 1.1 and 1.4 MPa to 1.8 and 2.0 MPa.

DISCUSSION

All methods of assessment indicate considerable improvement in the sands. The CPTU and the FDPM indicate improvement in the clayey silts but the shear wave velocity does not show any conclusive improvement. The use of penetration resistance alone does not provide much insight as to the likely change in stress-strain properties induced by ground improvement. The changes in the cyclic resistance of the clayey silts after improvement are unknown although the undrained shear strength appears to have increased. Both the CPTU and the PM indicate this. The PM also indicates an increase in stiffness of the clayey silt and the interpretation of the CPTU indicates an increased level of OCR. The differences in measured CPTU values are, however, close to the resolution of the instrument and must be interpreted with caution.

Ground improvement changes the density, but also the stress regime and the stress history of the deposit. Interpretation of in situ testing results should also consider stress level and stress history effects. Further insight into the effects of the ground improvement can be gained by measuring the degree of improvement by three different tests in the same hole. Chamber test data (Houlsby and Hitchman, 1988; Bellotti et al., 1985) have indicated that the CPT tip resistance in uncemented, unaged sands is a function of density, and horizontal (σ_h') and vertical (σ_v') effective stresses with the lateral stress being dominant. For example, the relationship between relative density (D_r), σ_h' and q_c for chamber tests on unaged Ticino Sand is given by Baldi et al. (1986) to be:

$$q_c = 248 \, \sigma_h'^{0.55} \exp[2.38 D_r] \tag{3}$$

where q_c is dependent on $(K_o)^{0.55}$ with $K_o = (\sigma_h'/\sigma_v')$. Such a relationship is commonly used for derivation of D_r from CPTU data in other moderately compressible, normally consolidated young sands such as those of the Fraser Delta.

Shear wave velocity (V_s) has also been found to vary with density (or void ratio) and stress level. Bellotti et al. (1996) found that the relationship between V_s, void ratio (e), and effective stress was given by an expression:

$$V_s = C_s \, [F(e)]^{0.5} (\sigma_a')^{na} (\sigma_b')^{nb} \tag{4}$$

where C_s is a function of grain characteristics
        $F(e)$ is a function of void ratio
        (σ_a') is the effective stress in the direction of particle motion
        (σ_b') is the effective stress in the direction of propagation
        and na and nb are exponents.

For Ticino sand, Bellotti et al. (1996) found C_s to be around 85 and na=nb=0.122. For the SCPTU case where waves propagate in an approximately vertical direction, (σ_a') is approximately equivalent to (σ_h') and (σ_b') to (σ_v'). If the expression is written in terms of K_o, the expression obtained is:

$$V_s = 85 \, [F(e)]^{0.5} \, (\sigma_v')^{0.244} K_o^{0.122} \tag{4a}$$

The term $K_o^{0.122}$ varies from 0.89 to 1.14 for the corresponding range: $0.4 < K_o < 3.0$. Similarly, $(K_o)^{0.55}$ ranges from 0.6 to 1.82 for the same range of K_o, and so the relationships predict that K_o has a small effect on V_s and a much greater effect on q_c.

Both q_c and V_s are different functions of the same parameters and it might be considered that a combination of the two measurements could provide information about the lateral stress and the relative density of the deposit.

The use of combinations of V_s and q_c

The results at the first site indicated that the initial q_c and V_s values at a depth of about 8.5 m were about 11 MPa and 175 m/s respectively and became about 20 MPa and 200 m/s after ground improvement. Using conventional interpretation procedures based on tip resistance (i.e. ignoring K_o), this would represent an increase in D_r from about 70% to 98%. It can be shown, however, that the after treatment values of tip resistance and V_s can also be attained by an increase in K_o from 0.45 to 1.4 and no increase in D_r. It is more likely that both lateral stress and D_r have been changed by the ground improvement and that the increase in tip resistance is due to a combination of the two. The relative effects are difficult to quantify.

Previous investigators have investigated the use of the ratio of G_o/q_c in the interpretation of SCPTU data. Baldi et al. (1986) note that G_o/q_c should go down with increasing D_r increases as q_c increases much faster with increasing D_r than does G_o. The data in Figures 2 and 3 show that this is indeed the case within the sand layer. More recently, Eslaamizaad and Robertson (1996) combined expressions for q_c and G_o and used calibration test data to derive by statistical correlation a relationship for the evaluation of K_o in which the effect of D_r has been eliminated. They gave the following relationship for the evaluation of K_o:

$$G_o/p_a = 334.90(q_c'/p_a)^{0.25}(\sigma_v'/p_a)^{0.332} K_o^{0.462} \tag{5}$$

This can be rearranged to give an expression for K_o as follows:

$$K_o = 3.4 \times 10^{(-6)} (p_a/\sigma_v')^{0.718}[(G_o/p_a)/(q_c'/p_a)^{0.25}]^{(2.165)} \tag{6}$$

If it is assumed that for any particular depth σ_v' remains constant during densification, the change in K_o or lateral stress induced by ground improvement can be estimated from this expression. An increase in K_o should result in an increase in the parameter, $[(G_o/p_a)/(q_c'/p_a)^{0.25}]$. In Figures 2 and 4, it can be seen that, in general, it has increased within the sands as a result of vibro-replacement. The estimated values of K_o before and after ground improvement indicate a significant increase in lateral stress. If G_o calculated from Equation (2) is substituted in Equation (6), the resulting expression includes the term $(V_s)^{4.33}$. The estimated K_o values will thus be very sensitive to errors in V_s. There is still a good indication that an increase in lateral stress has occurred.

Although steps were taken to minimize the interaction between the stone columns and the beam used to transmit the shear waves, the observed increase in V_s could be due to the shear waves travelling primarily through the stone columns. Baez (1995) measured values of V_s ranging from 275 and 396 m/s at depths of between 5 to 7 m in stone columns constructed by vibro-replacement in a poorly graded fine to medium sand. These values compare to measured V_s values for improved ground of around 200 to 225 m/s. If the shear waves were travelling predominantly through stone columns, the calculated shear wave velocity for each depth increment should reflect the much faster stone column V_s value. This is not consistent with the values observed, lending some confidence to the assumption that the measured V_s corresponds to the improved sand.

The important issue of ageing of the sand has not been included in this discussion. In fact, both q_c and V_s have been observed to decrease in sands when measured soon after ground treatment and to then show an increase with time after treatment (Schmertmann, 1991). This aspect of the response of soils to densification is poorly understood and requires further research.

Implications for Design

The above discussion suggests that both the density and the stress regime have been changed by ground treatment. For design of foundations to be supported in or on the improved ground, the stress deformation characteristics under both static and cyclic loading conditions will govern the anticipated soil-structure interaction. It would be desirable that the techniques used for quality control should provide an unambiguous determination of the stress-deformation properties of the improved ground. Jamiolkowski and Pasqualini (1992) have indicated that the degree of overconsolidation has much more effect on the stiffness of the soil than any increase in density. It would thus be preferable to carry out tests to separate the effects of increased density from the effects of increased OCR or stiffness. This would allow the technical specification for ground improvement to be written in terms of achieved stiffness and strength.

Neither the CPTU nor V_s measurement taken on their own give reliable stiffness information for foundation design, because the effect of the ambient lateral stress on the former cannot be assessed and the latter allows calculation of only a small strain stiffness. The pressuremeter expansion curve, however, provides a measured load-deformation curve for the treated soil and foundation design using the results of Menard pressuremeter testing has proved effective on sites which have been improved (Debats et al., 1998). Methods of assessing the stress-strain behaviour of improved ground under cyclic loading are less well developed. Design procedures developed for the Menard pressuremeter are based on empirical relationships to foundation performance which take into account the soil disturbance caused by pressuremeter installation. It is likely that similar rules could be developed for the FDPM or cone pressuremeter. The seismic cone pressuremeter could then offer three different measures of the improvement obtained and procedures similar to those above could be followed to use the information obtained. This requires further research.

Other Indicators of Improvement

The previous section advocates the use of a combination of in situ tests to provide insight to the stress-deformation behaviour of the treated soil. Resistivity measurements obtained at the second site and at others have suggested the use of such measurements as a further indicator of ground improvement (Daniel et al., 1999).

Current approaches to interpretation of resistivity data use Archie's Law. Archie (1942) proposed the following empirical relationship to describe the effects of pore fluid resistivity and porosity (a measure of soil density) on bulk soil resistivity in saturated soils:

$$\rho_b = \rho_f \cdot a \cdot n^{-m} \tag{7}$$

where: ρ_b = bulk soil resistivity (ohm·m);
ρ_f = pore fluid resistivity (ohm·m);
n = soil porosity (void volume / total volume); and
a, m = local soil constants.

If it is assumed that the soil constants do not change during densification the following relationship can be written (where (') indicates post-densification values):

$$\rho_b'/\rho_b = (\rho_f'/\rho_f) \cdot (n'/n)^{-m} \tag{8}$$

or

$$n'/n = ((\rho_b'/\rho_b) \cdot (\rho_f/\rho_f'))^{-1/m} \tag{9}$$

Published values of the soil constant (m) range from 1.3 to 2.2 for granular soils with non-conductive grains (Archie, 1942, Hyde and Hunter, 1998). Archie originally correlated the (m) constant to soil consolidation but it is often assumed to be a function of grain shape.

Equation (9) can be used with RCPTU measurements, groundwater resistivity measurements and an assumed value of (m) to estimate the ratio of post to pre densification porosity. Figure 7 shows the pre and post-densification resistivity data from the second site, including the calculated porosity ratio assuming values of 1.3 and 2.2 for (m). Water sampling was not performed at this site. However, water samples were obtained at another site before and after densification using the UBC modified BAT groundwater sampler (Campanella et al. 1995) to investigate whether the change in resistivity was due to changes in groundwater chemistry during the vibro-replacement. No effect was observed. Consequently, the fluid resistivity ratio was assumed equal to one. The pre-densification resistivity is likely higher than post densification between 0 and 4 meters depth because of partial saturation of the fill sands prior to densification.

The porosity ratio profiles for both (m) values within the Fraser River sands are above the minimum possible porosity ratio of 0.79 calculated from representative values of minimum and maximum void ratios. The results suggest that the porosity of the sand adjacent to the resistivity module is lower as a result of the ground improvement. This zone of soil is in a region of large disturbance during cone penetration with the magnitude and distribution of volumetric strain varying with density and stress level, both of which are increased by ground improvement. This observation of changed resistivity has been confirmed at other vibro-replacement sites but requires further research to explain how the observed phenomenon relates to the change in soil stress-strain behaviour induced by vibro-replacement.

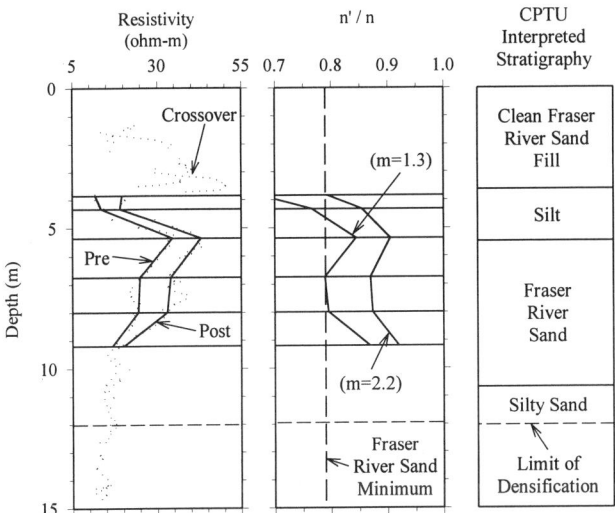

Figure 7 - Pre- and Post-Densification Resistivity Data and Calculated Porosity Ratio from a Densification Site

CONCLUSIONS

Liquefaction assessment, the need for ground improvement and the quality control and assurance of the ground treatment are conventionally based on the interpretation of penetration testing. More recently, shear wave velocities have also been used.

When assessing the degree of improvement achieved by ground treatment, it is important to consider the effects of both increases in density and of lateral stress on in situ test results. The results presented suggest that the additional information provided by a combination of tests is likely to improve our understanding of the changes in soil behaviour achieved by ground treatment. The ratio of G_o from V_s measurements to tip resistance appears to provide information about changes in lateral stress. Pressuremeter testing seems to be more sensitive to improvement in silts than either penetration resistance or shear wave velocity. Due to soil variability, all parameters should be measured in the same sounding. The seismic cone pressuremeter or other combinations of in situ tests are worthy of further research.

The addition of bulk soil electrical resistivity to the suite of data collected during piezocone penetration tests is a simple, inexpensive way to increase the body of data that can be used to assess the effect of vibro-replacement. Resistivity measurements also have potential for clarifying the relative contributions of increases in lateral stress and density to increased penetration resistance. The resistivity data generally yield the same trends in density as the conventional in situ testing. Additional research is required to estimate the magnitude of volumetric soil strain during piezocone penetration and the effect of this strain on resistivity measurements. This parameter may also prove useful as an indication of changes in soil response induced by ground improvement.

Further research is also required into the assessment of the behaviour under cyclic loading of fine-grained soils and particularly liquefiable silts. Although vibro-replacement appears to be capable of increasing the strength and stiffness of such soils as indicated by the results of PM testing and CPTU testing, the degree of improvement attained and its influence on the cyclic resistance of silts requires further research.

ACKNOWLEDGMENTS

The authors acknowledge the financial support of the Natural Sciences and Engineering Research Council of Canada and the GREAT Award program of the B.C. Science Council. They also acknowledge the support of the University of British Columbia Civil Engineering Department, Geopac West Ltd. and Geopacific Consultants Ltd. Technicians Scott Jackson and Harald Schrempp built and maintained the equipment.

REFERENCES

Archie, G.E., (1942). "The Electrical Resistivity Log as an Aid in Determining Some Reservoir Characteristics." *Transactions of American Institute of Mining Engineers*, Vol. 146, pp.54-62.

Armstrong, J.E. (1981). "Post-Vashon Wisconsinan glaciation, Fraser Lowland, British Columbia." *Bulletin 322*, Geological Survey of Canada, 34 p.

Baez, J.I. (1995). "A Design Model for the Reduction of Soil Liquefaction by Vibro-Stone Columns". *PhD Dissertation*, Dept. of Civil Engineering, University of Southern California.

Baldi, G., Bellotti, R., Ghionna, V.N., Jamiolkowski, M. and Pasqualini, E. (1986). "Interpretation of CPT's and CPTU's Second Part: Drained Penetration of Sands." *Proceedings of 4th International Geotechnical Seminar on Field Instrumentation and In Situ Measurements*, Nanyang Technical Institute, Singapore, 129-156.

Bellotti, R., Crippa, V., Pedroni, S., Baldi, G., Fretti, C., Ostricati, D., Ghionna, V., Jamiolkowski, M. and Pasqualini, E. (1985). "Laboratory validation of in-situ tests." *Italian Geotech. Assoc. (AGI) Jubilee Volume, Proceedings*, 11th International Conference on Soil Mechanics and Foundation Engineering, San Francisco, A.A. Balkema Publishers, Rotterdam.

Bellotti, R., Jamiolkowski, M., Lo Presti, D.C.F. and O'Neill, D.A. (1996) "Anisotropy of small strain stiffness in Ticino sand." *Geotechnique*, 46, No.1, 115-131.

Campanella, R.G., Gillespie and Robertson, P.K. (1982). "Pore pressures during cone penetration testing. ESOPT-II, *Proceedings of the Second European Symposium on Penetration Testing*, Amsterdam. Vol. 2, A.A. Balkema, Rotterdam, 507-512.

Campanella, R.G., Martens, S., Tomlinson, S. and Davies, M.P. (1995). "In-Situ measurement of hydraulic conductivity in sands." *Proceedings of 48th Canadian Geotechnical Conference*, Canadian Geotechnical Society, Vancouver, BC, Vol. 1, 309-318.

Daniel, C.R., Giacheti, H.L., Howie, J.A. and Campanella, R.G. (1999). "Resistivity Piezocone (RCPTU) Data interpretation and Potential Applications". *Proceedings*, *XI Panamerican Conference on Soil Mechanics and Geotechnical Engineering*, Vol. 1, Foz do Iguassu, Brazil, 361-369.

Debats, J-M.,M., Frank, R.A., Gambin, M.P., and Savasta, P.A.(1998). "The Menard pressuremeter for quality control of soil densification." *Geotechnical Site Characterization*, P.K. Robertson and P.W. Mayne (eds), Vol. 2, A.A. Balkema, Rotterdam, 765-770.

Eslaamizaad, S. and Robertson, P.K. (1996). "Estimation of in situ lateral stress and stress history in sands." *Proceedings of 49th Canadian Geotechnical Conference*, St. John's, Newfoundland.

Houlsby, G.T., and Hitchman,R. (1988). "Calibration chamber tests of a cone penetrometer in sands." *Geotechnique*, Vol. 38 (1), 39-44.

Hyde, C.S.B. and Hunter, J.A. (1998). "Formation Electrical Conductivity-Porewater Salinity Relationships in Quaternary Sediments from Two Canadian Sites." *Proceedings,* Symposium on the Application of Geophysics to Engineering and Environmental Problems (SAGEEP), Chicago, Illinois, 499-510.

Jamiolkowski, M. and Pasqualini, E. (1992). "Compaction of Granular Soils – Remarks on Quality Control." *Soil Improvement and Geosynthetics*, R.H. Borden, R.D. Holtz and I. Juran, (Eds.), Geotechnical Special Publication No. 30, v. 2., ASCE, New York, 902-914.

Mitchell, J.K., Cooke, H.G., and Schaeffer, J.A. (1998). "Design Considerations in Ground Improvement for Seismic Risk Mitigation." *Geotechnical Earthquake Engineering and Soil Dynamics III*, Vol. 1, Geotechnical Special Publication No.75, ASCE, Reston, Virginia, 580-613.

NCEER (1997). Summary Report, *Proceedings of the NCEER Workshop on the Evaluation of Liquefaction Resistance of Soils*. Technical Report 84602-4081, Youd, T.L. and Idriss, I.M. (eds.), National Center for Earthquake Engineering Research, State Univ. of New York at Buffalo, 276 p.

Robertson, P.K., and Wride, (1997). "Cyclic Liquefaction and its Evaluation based on the SPT and CPT." *Proceedings of the NCEER Workshop on the Evaluation of Liquefaction Resistance of Soils*. Technical Report 84602-4081, Youd, T.L. and Idriss, I.M. (eds.), National Center for Earthquake Engineering Research, State Univ. of New York at Buffalo, 41-88.

Robertson, P.K., Campanella, R.G., Gillespie, D. and Rice, A. (1986). "Seismic CPT to measure in-situ shear wave velocity." *Journal of Geotechnical Engineering*, ASCE, 112 (8): 791-804.

Schmertmann, J.H. (1991). "The mechanical aging of soils." 25th Karl Terzaghi Lecture, *Journal of Geotechnical Engineering*, ASCE, 117(9):1288 - 1330.

Wang, W. (1979). "Some findings in Soil Liquefaction." *Water Conservancy and Hydroelectric Power Scientific Research*, Beijing, China.

Wroth, C.P. (1984). "The interpretation of in-situ soil tests." 24th Rankine Lecture, *Géotechnique*, 34 (4): 449-489.

EFFECT OF PILE INSTALLATION ON STATIC AND DYNAMIC PROPERTIES OF SOFT CLAYS

Christopher E. Hunt Student M. ASCE, Juan M. Pestana M. ASCE,
Jonathan D. Bray M. ASCE, and Michael F. Riemer[1] Associate M. ASCE

ABSTRACT

Static and dynamic soil properties were monitored around a closed-ended steel pipe pile driven in a deep deposit of San Francisco Young Bay Mud. Shear wave velocity measurements were taken prior to pile driving to establish the baseline condition and at selected times after pile driving to document the changes as a function of time. Additional instrumentation and monitoring activities included pore water pressure response and lateral soil deformations as a result of pile driving. A series of conventional laboratory tests were performed on samples collected prior to pile driving and 8 months afterwards. Field measurements after pile driving indicate a significant increase of shear wave velocity as a function of time accompanied by excess pore pressure dissipation. Laboratory tests show a significant increase in the strain to failure with slight increases in undrained strength.

INTRODUCTION

Over the last several decades a large body of research has been performed to evaluate the changes occurring in the soil properties surrounding driven piles. Numerous projects using model scale field piles, many of them extensively instrumented, have examined loading conditions, total stresses and pore pressures on and around the piles (e.g. Reese and Seed, 1955; Seed and Reese, 1957; Lo and Stermac, 1965; Torstensson, 1973; Bogard and Matlock, 1990; Bond and Jardine, 1991; Lehane and Jardine, 1994). Several projects have used full-scale piles and field and/or pile instrumentation to study the same features at larger scale in a variety of environments (e.g. Orrje and Broms, 1967; Airhart et al., 1969; Roy et al., 1981; O'Neill et al., 1982; Soares and Dias, 1989). Most of this research has focused on the evaluation of side resistance (i.e., skin friction) and has been motivated by the need to more accurately predict pile capacities for

---

[1] Graduate Student Researcher, Asst. Professor, Professor, and Asst. Adjunct. Professor respectively, Dept. of Civil and Environmental Eng., University of California, Berkeley, California 94720.

offshore drilling platforms and foundations for major highway bridges under primarily monotonic or cyclic loading (e.g., O'Neill, 1999).

More recently, there has been a renewed effort in seismic soil-pile-structure interaction testing and analysis to provide guidelines for foundation design and the evaluation of retrofit measures for large bridges founded on soft clay deposits. To support such research, there is a need for well-documented case histories reporting changes in the static and dynamic properties of soft clays within the zone of influence resulting from pile installation (e.g., Pestana, 1995). This paper focuses on characterizing the soil's response to the installation of a full scale (61 cm diameter) steel pipe pile driven closed ended through a deep deposit of soft clay. In contrast to previous research efforts, focus is placed on shear wave velocity (v_s) measurements and stress-strain properties of the clay surrounding the pile as a function of time after installation. Measurement of these properties is of paramount importance in the evaluation of the stiffness of the soil-pile foundation system required for seismic soil-pile-superstructure analysis procedures. In addition, shear wave velocity measurements, in combination with values of the pore pressures within the soil, can lead to a more complete understanding of the state of stresses in the soil within the area of influence of the pile. The combination of field instrumentation and in-situ and laboratory testing provides a comprehensive view of the soil over the lifetime of the research project, from pre- to post-driving conditions.

SITE SELECTION

The California Department of Transportation (CALTRANS) provided access to a suitable site south of San Francisco near Islais Creek, with the added advantage that a pile load testing program had been performed only a few hundred feet away in 1993 (CALTRANS and DFI, 1993), providing extensive information on the subsurface conditions. In the late 1800's the site was a tidal salt marsh where the clay had been deposited over thousands of years, intersected by numerous stream channels. According to.Radbruch and Schlocker (1958) fill was placed throughout the region at various times from about 1890 to 1940. Figure 1 shows a general soil profile of the site, consisting of approximately 4.5 m of miscellaneous fill containing rubble and occasional cavities, 11 m of soft Young Bay Mud, 1.5 m of clayey sand, and another 16.5 m of Young Bay Mud, underlain by a significantly stiffer clayey sand deposit at a depth of approximately 32 m. Joint interpretation with previous soil investigation reports shows a consistent areal extent with essentially horizontal layering within the domain of interest.

SITE INVESTIGATION AND INSTRUMENTATION PLAN

Several steps were taken to characterize the site conditions in the immediate vicinity of the planned pile installation. Seismic cone penetrometer tests with pore pressure measurements (SCPT-U) were performed at distances of 7.3 m from the pile location, to a depth of 34 m each. These tests provided tip and frictional resistances, as well as a penetration pore pressure profile (Figure 1), and baseline shear wave velocity

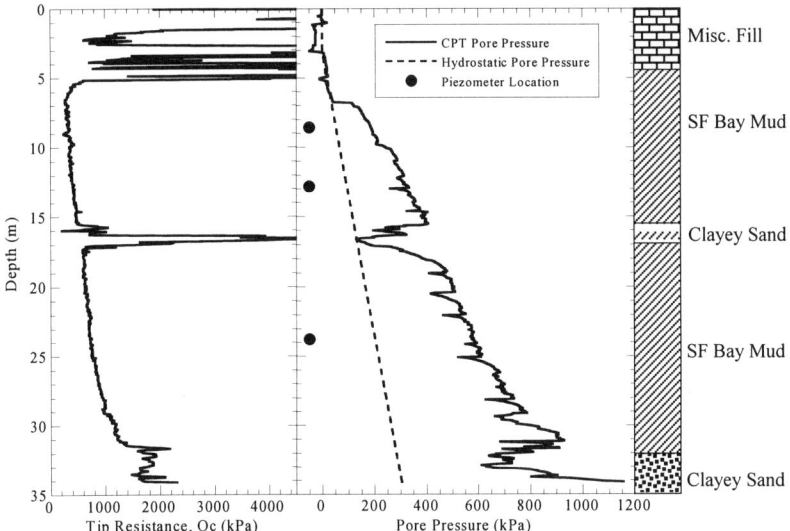

Figure 1: Soil profile and CPT measurements

measurements. In addition, pore pressure dissipation tests were performed at 9.1 m and 18.3 m, thus providing estimates of the rate of dissipation of excess pore pressures caused by cone penetration.

Ten pneumatic piezometers were placed in boreholes, spaced at three radial distances from the pile and at three depths, with one piezometer placed at greater distance from the pile to measure local water table conditions. Pneumatic devices were selected in order to ensure survival after pile installation. The selected piezometer depths (8.5 m, 12.8 m, and 23.8 m) correspond to three locations of fairly uniform soil conditions.

Figure 2 shows the location of borings relative to the pile. Borings 1, 2 and 3 contain two piezometers each, at depths of 8.5 m and 12.8 m. Borings 4, 5 and 6 have one piezometer each at 23.8 m, with inclinometer casing grouted in place above 22.4 m. Boring 7, approximately 8 pile diameters from the wall of the pile, contains one piezometer at 6.7 m. Each piezometer was placed within a sand cell approximately 1 m in height with a bentonite seal above to ensure localized measurements. The entire space between the two sand cells in borings 1, 2 and 3 was filled with a bentonite plug. The hollow circles shown on Figure 2 correspond to the measured bottom of the inclinometer casings in borings 4, 5 and 6 prior to pile installation, and indicate the deviation from the intended vertical alignment (one to two percent along the 22.4 m cased length). Boring 9 was not part of the initial investigation, having been drilled 8 months after pile installation in order to obtain post-pile samples and shear wave velocities adjacent to the pile wall. All borings were performed with rotary wash drilling methods.

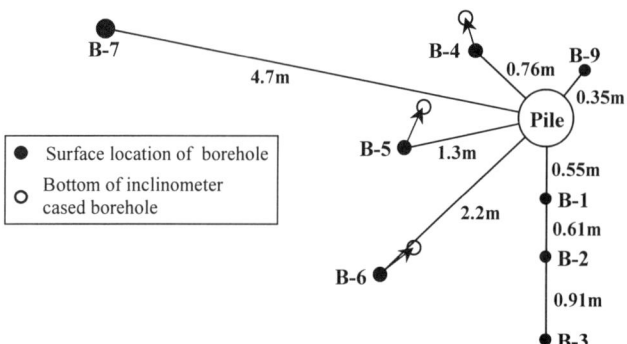

Figure 2: Borehole layout relative to pile

Inclinometer casing in borings 4, 5 and 6 were assumed locked within the 4.5 m of fill at the surface. The pile was isolated from the fill by predrilling and installing a 76 cm diameter steel casing to a depth of 5.2 m prior to driving. This not only prevented displacement of the inclinometer casings, but also prevented the pile from pushing fill material into the Bay Mud below. Accurate surface surveys of the top of all inclinometer casings prior to and post driving showed negligible displacement, allowing them to be used as reference points for deformation measurements.

After completion of installation of all instrumentation, a period of two months was provided for equilibration of piezometers and for the grout to set up completely around the inclinometer casings, as well as to allow the clay to recover from the extensive drilling operations. Baseline readings were then taken on September 30 and October 1, 1997, and the pile was driven October 2, 1997.

INSTRUMENTATION MONITORING

The pile driven at the site was a 61 cm diameter, closed-end steel pipe pile. It was driven 36 m into the ground, fully penetrating both layers of Bay Mud, tipping out in the stiffer clayey sand layer below. Measurements were made with a level in between pile driver blows, and adjustments were made as necessary, to ensure a vertical pile alignment. Pore pressures induced by pile installation were first measured within hours after driving. Inclinometer and shear wave velocity measurements began on the following day. Preliminary results and interpretation of shear wave velocity measurements were presented by Hunt et al. (1998). Additional data and a detailed analysis of the records have yielded a different shear wave velocity interpretation, which is intended to supersede the previously reported results (Hunt et al, 2000).

Pore Pressures

Figure 3 shows pore pressure measurements from the two piezometers in B-1 and the single piezometer at the bottom of B-4, which correspond to 8.5 m, 12.8 m, and 23.8 m depths. Readings are plotted as a percent dissipation of the excess pore pressures measured immediately after pile driving. As expected, the pore pressures dissipate at a decreasing rate as the clay undergoes radial and vertical consolidation. Except for one erratic measurement, the piezometer at a depth of 12.8 m displays the quickest dissipation, likely due to its proximity to the sandy layer below. Pore pressures generated immediately after pile driving (Δu_{max}) and their ratio with the pre-pile vertical effective stresses (σ'_{vi}) at the three depths are also shown on the figure. The shallow piezometers in B-1 produce the largest relative pore pressure generation ($\Delta u_{max}/\sigma'_{vi}$), while B-4 shows a smaller relative increase. These results are consistent with the fact that the piezometers in B-4 are nominally farther away from the pile (Figure 2). Hunt et al. (2000) present full details of measured pore pressure dissipation.

Lateral Deformations

Figure 4 shows lateral displacements inferred from inclinometer measurements for Boring 5 along a radial axis towards the center of the pile. Deflections shown are relative to the baseline measurements taken prior to driving the pile. The negligible deformation shown in the upper 4.5 m supports the hypothesis that drilling through and casing the fill prior to pile driving allowed the inclinometer casing to be locked in the top fill material. However, large deformations occurred in the soil at the depth where pile driving began. The dashed line shows the relative displacements predicted by the

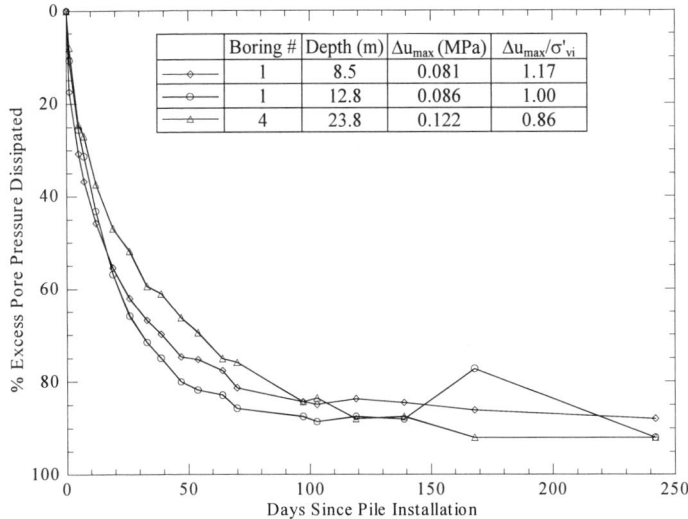

Boring #	Depth (m)	Δu_{max} (MPa)	$\Delta u_{max}/\sigma'_{vi}$
1	8.5	0.081	1.17
1	12.8	0.086	1.00
4	23.8	0.122	0.86

Figure 3: Excess pore pressure dissipation

cylindrical cavity expansion theory assuming undrained soil response (e.g., Gibson, 1950; Vesic, 1972; Randolph et al., 1979). These predictions appear to provide a good estimate of the pile-induced lateral deformations in Young Bay Mud. Discrepancies between predicted and observed deflections directly below the fill are likely due to slight adjustments made to the driving alignment of the pile during the first few meters, when it lacked sufficient embedment to follow the vertical path required.

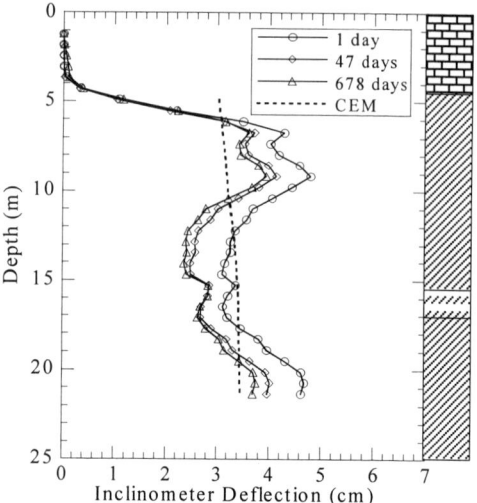

Figure 4: Inclinometer data for B-5

The effects of pore pressure dissipation and radial consolidation are apparent in the deflection measurements at 47 and 678 days after pile installation, which indicate displacement back towards the pile. While movement within the clayey sand layer appears complete after 47 days, in the Young Bay Mud, where pore pressure dissipation at that time is in the 70% to 80% range, some small deflection occurs between 47 and 678 days.

Shear Wave Velocities

The OYO suspension logging method (Nigbor and Imai, 1994) was used to measure shear wave velocities in each of the three inclinometer casings. The key components of the measurement device are a source mechanism, and two geophones which are separated by a known spacing (typically 1 m). Triggering of the source produces a signal that is recorded independently by the two geophones. The shear wave velocity is then calculated from the difference in shear wave arrival times in the two geophones, and thus represents the average shear wave velocity over that separation.

The suspension logging method was selected for this research project for two reasons. First, the shear wave velocities measured are averaged over a relatively small volume of soil centered on a single borehole, as opposed to a crosshole method in which the source and geophone are in separate boreholes and the velocities are averaged over the entire intervening distance. These localized measurements provide invaluable information on the variation in dynamic properties as a function of the radial distance from the pile. Second, as the source and geophones are all on one device, the process for measuring in a borehole is relatively efficient. The tool is lowered to the bottom where the first set of measurements is made, and is then raised incrementally, collecting readings as often as necessary to achieve a detailed profile. All three borehole profiles can be carefully measured in a few hours.

MEASURED SIGNAL WAVEFORMS

While suspension logging requires two geophones in order to measure the shear wave velocity over an incremental distance, some insight into this particular field experiment can be gained from looking at the received records of only one geophone over time. Figure 5 shows the lower geophone records in Boring 5 at a depth of 13 meters for each series of measurements after pile installation. At each date, the raw record is shown (thin line) overlain by a filtered version (thick line). The source and receivers used on each of the dates shown were the same, and while there can be some slight discrepancy between the triggering of the source and the beginning of the recorded signal, it is typically small and would not affect the overall trends apparent in the figure.

The dashed line on Figure 5 shows one possible interpretation of the records, which is a gradual decrease in the travel time of the shear wave, which in turn corresponds to an increase in the shear wave velocity of the soil over time. These increases are primarily related to the dissipation of generated pore pressures and the resulting consolidation of the clay. While the pre-pile waveforms are in general not comparable with the post-pile waveforms, and thus are not shown on the figure, velocity comparisons indicate an initial decrease in velocity close to the pile immediately after driving and a subsequent increase over time.

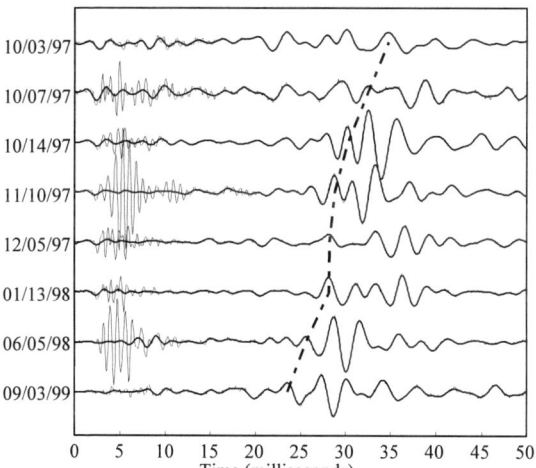

Figure 5: Time variation in near geophone record

MEASURED SHEAR WAVE VELOCITIES

Figure 6a shows the final shear wave velocity (v_s) profiles for borings 4 through 6 twenty-three months (701 days) after driving the pile (numbers in parentheses are average distances from the pile wall). These values are compared to the baseline v_s profile obtained from Boring 5 prior to driving the pile. In the depth range from 9 m to 14 m, which lies completely within the clay, baseline v_s is on average 90 m/sec. Over this same range, the average v_s measurements twenty-three months later are 140 m/sec, 120 m/sec, and 100 m/sec for borings 4, 5, and 6 respectively. These measurements indicate a significant increase in v_s near the pile, with diminishing effect with increasing distance from the pile. These observations are qualitatively in agreement with pore pressure measurements. The soil closest to the pile experiences the largest deformations and the largest excess pore pressure generation, with diminishing effect with increasing distance from the pile face.

Also of interest is the significant change in v_s in the clayey sand layer that runs from 15.5 m to 17 m, which increases from approximately 130 m/sec to a maximum of 255 m/sec at 16.5 m. Similar values of v_s for borings 4 and 5 in the 16.5 m to 18.5 m range are expected. Boring 4 in fact angles away from the pile while Boring 5 angles towards the pile, resulting in both borings being approximately 1 m from the pile wall at a depth of 18 m, as opposed to the planned 0.76 m and 1.3 m which occur at the surface for borings 4 and 5 respectively. These results therefore support the consistency of field measurements.

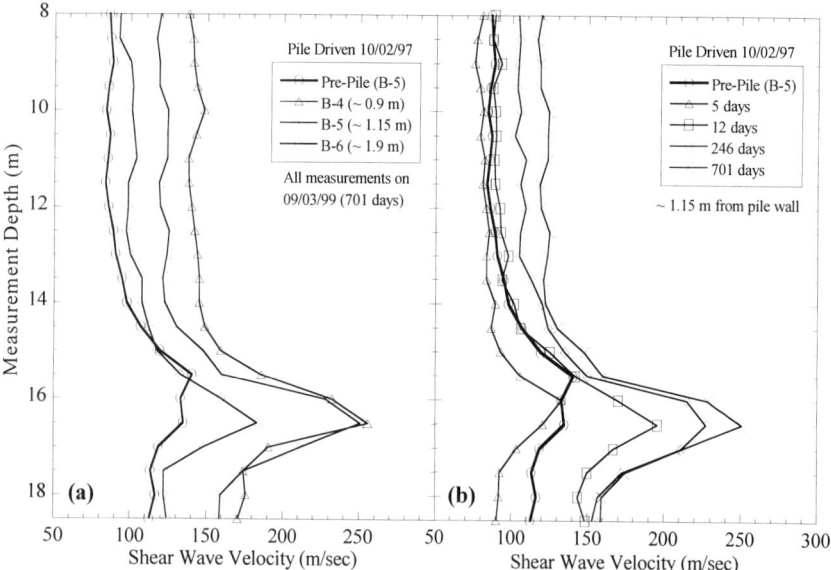

Figure 6: Shear wave velocity variations (a) with distance from pile (701 days after driving) and (b) in Boring 5 over time after driving

Figure 6b shows the evolution of v_s over time for Boring 5. Attempts at interpreting suspension logging records within all three inclinometer casings were unsuccessful immediately after pile driving, likely due to the extremely high initial pore pressures in the soil. By five days after installation roughly 30% of the pore pressures generated in Boring 5 had dissipated and the geophone records were coherent enough to develop a velocity profile. The average baseline (pre-pile) v_s in the range from 9 m to 14 m in Boring 5 is 90 m/sec. Over this same depth range six days after driving, the average v_s is 80 m/sec. The corresponding velocities 12, 246, and 701 days after pile driving are 90 m/sec, 110 m/sec, and 120 m/sec respectively. Thus, there is an initial decrease after driving due to the increased pore pressures, which correspond to a reduced effective stress state in the soil, followed by a gradual increase beyond the baseline shear wave velocities as the pore pressures dissipate and the soil consolidates. Additional v_s increases after near complete pore pressure dissipation ($\sim$ 246 days) are attributed to aging effects during secondary compression. These late stage increases are not unusual, and have been witnessed by numerous researchers (e.g., Humphries and Wahls, 1968; Afifi and Woods, 1971; Stokoe and Richart, 1973; Anderson and Woods, 1976) studying time and pressure effects on shear wave velocity and dynamic shear modulus, primarily using resonant column testing devices.

A shear wave velocity profile was also obtained from Boring 9, drilled adjacent to the pile wall eight months after pile driving. The average v_s over the depth range of 9 m to 14 m in this boring was 65 m/s, over 25% lower than the pre-pile velocity. The most probable explanation for this departure from the trends in B-4 through B-6 is that the soil in the immediate vicinity of the pile was significantly remolded by the driving process. Thus, potential increases in velocity from an increase in density and stress were offset by the remolding of the clay structure and any bonds that had developed over its post-depositional history. Additionally, significant excess pore pressures may remain in the clay adjacent to the pile at the time of measurement, although at greater distance excess pore pressures were nearly completely dissipated.

LABORATORY TESTING

An extensive laboratory testing program was established to measure relevant properties, and included soil specimens collected during installation of instrumentation prior to pile driving and eight months afterwards in Boring 9. Testing included standard index tests (e.g. water content, Atterberg limits) as well as anisotropically consolidated, undrained triaxial compression tests, constant rate of strain (CRS) consolidation tests to large stresses, triaxial consolidation tests with imposed radial drainage, and cyclic simple shear tests.

Tests on pre-pile specimens indicated that the Young Bay Mud was a highly plastic clay (PI $\approx$ 44 to 59) and lightly overconsolidated (OCR $\approx$ 1.2 to 1.4). Ko was estimated to be 0.62. Figure 7 shows the water contents and unit weights measured prior to and after pile installation. As expected, a decrease in water content and an increase in unit

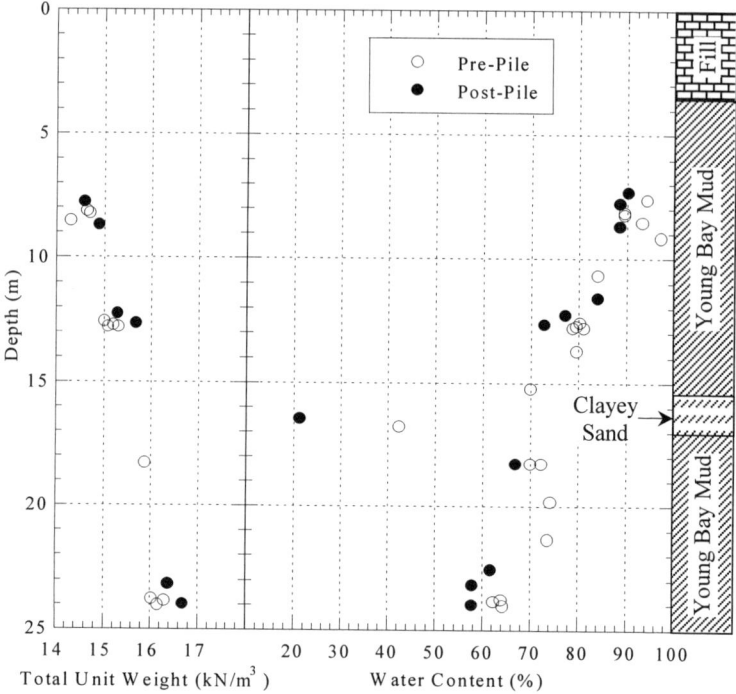

Figure 7: Pre- and post-pile water content and unit weight values

weights occurred as the pore pressures dissipated radially away from the pile and the soil consolidated back towards the pile.

Figure 8 shows standard e-log σ'_v compression curves from CRS consolidation tests on specimens from depths of 8.5 m, 12.8 m, and 23.8 m. The pile-induced remolding of the soil has led to an apparent loss of memory in the post-pile specimens, as the transition from the overconsolidated to normally consolidated regime is more gradual than in the pre-pile specimen. Given the gentle curvature of the post-pile test results, it is difficult to estimate a maximum past pressure after the pile has been driven.

Figure 9 shows plots of deviatoric stress vs. axial strain from strain controlled triaxial compression tests on pre- and post-pile specimens taken from depths of 7.75 m, 12.4 m, and 23.25 m. All test specimens were consolidated anisotropically with the ratio σ'_h/σ'_v equal to 0.62 in an attempt to replicate the in-situ pre-pile stress conditions. While the post-pile specimen is admittedly no longer at the pre-pile stress conditions, tests were performed identically in order to facilitate comparison of the results. Data beyond 10% strain are indicated with dashed lines, since the stress-strain relationship begins to lose

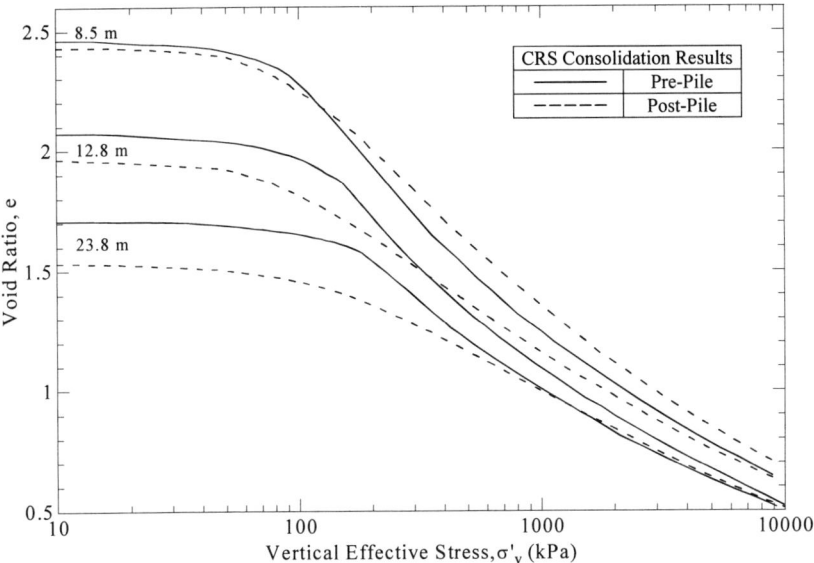

Figure 8: Constant rate of strain consolidation tests on pre- and post-pile specimens

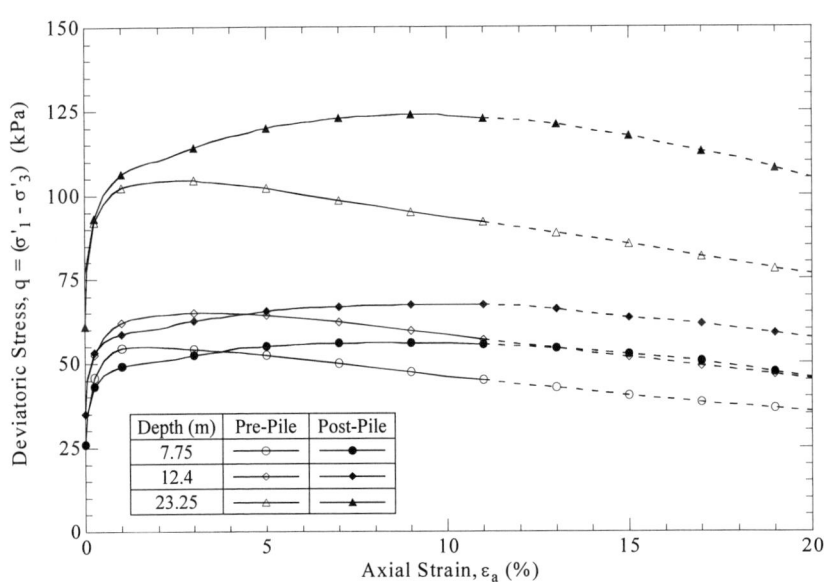

Figure 9: Deviatoric stress vs. axial strain for A-CU triaxial tests
consolidated to approximate pre-pile in-situ stresses

validity as the test specimen deviates from its assumed cylindrical shape. Results in the figure indicate minor increases in the peak strength of the two shallow specimens, with a larger increase for the deep specimen. However, soil ductility has increased significantly with the axial failure strain (i.e. the strain required to reach peak deviator stress) shifting from the 1.5% to 3% range to the 8% to 9% range. It is apparent that any strength gains that would be expected in the denser post-pile specimens from shallow depths have been offset by the breakdown in soil structure caused by pile installation.

CONCLUSIONS

This paper describes the instrumentation and measurements made to assess the effect of pile driving on static and dynamic properties of a soft clay (San Francisco Young Bay Mud). Due to pile installation, the soil transformed from a pre-installation state of equilibrium, to a transient deformed state with high values of pore pressure. Through subsequent consolidation, the clay is approaching a new state of equilibrium with higher values of shear wave velocity at distance from the pile, which translate to an increased stiffness of the soil. However, directly adjacent to the pile, shear wave velocities measured 8 months after pile installation were lower than the original baseline values.

Laboratory tests on pre- and post-pile specimens provide direct evidence of the consolidation processes, yielding specimens that are denser and lower in water content. The remolding caused by pile installation has essentially erased the prior stress history of the specimens as witnessed in consolidation tests. In addition, triaxial shearing tests indicated that while no significant strength gains had occurred, post-pile specimens were much more ductile, with failure strains three to four times higher than in pre-pile specimens.

Changes at the site are significant over a relatively large area of influence, with shear wave velocity variations at distances exceeding 3 pile diameters from the pile wall. These variations could have a significant impact on the seismic soil-pile foundation response of a pile group where typical separations can be 2 to 3 pile diameters (center to center). The results of this research have the potential to impact the seismic evaluation and design criteria used for new structures and to provide a more rational basis for seismic retrofit of existing foundation systems.

ACKNOWLEDGMENTS

This research was supported by the California Department of Transportation (CALTRANS) through contract No. RTA-59A130, by the NSF Career Award to the second author, and by a David and Lucille Packard Foundation Fellowship to the third author. The authors thank Robert Steller and GEOVision, for providing generous assistance with the collection of suspension logging data, and Robert Y. Chew Geotechnical Engineering, Inc. and Demetrious Koutsoftas for invaluable assistance with instrumentation selection, installation and monitoring.

REFERENCES

Afifi, S.S. and Woods, R.D. (1971) "Long-term pressure effects on shear modulus of soils," J. Soil Mech. Found. Div., ASCE, 97 (SM10), 1445-1460.

Airhart, T.P., Coyle H.M., Hirsch, T.J. and Buchanan, S.J. (1969) "Pile-Soil System Response in a Cohesive Soil," Performance of Deep Foundations, ASTM, pp. 264-294.

Anderson, D.G. and Woods, R.D. (1976) "Time-dependent increase in shear modulus of clay," J. Geot. Engrg., ASCE, 102 (GT5), 525-537.

Bogard, J.D. and Matlock, H. (1990) "In-Situ Pile Segment Model Experiments at Harvey, Louisiana," Proc., 22nd Annual Offshore Tech. Conf., Houston.

Bond, A.J. and Jardine, R.J. (1991) "Effects of installing displacement piles in a high OCR clay," Geotechnique, 41 (3), pp. 341-363.

California Department of Transportation (Caltrans), and Deep Foundations Institute (DFI), (1993) "CALTRANS Pile Load Test Results at a Deep Bay Mud Site Using Various Pile Types," Proc., 1993 DFI and Caltrans Pile Load Test Seminar.

Gibson, R.E., Oct. 1950, "Discussion on paper by Guthlac Wilson," Journal of the Institution of Civil Engineers, 34 (8), 382-383.

Humphries, W.K. and Wahls, H.E. (1968) "Stress history effects on dynamic modulus of clay," J. Soil Mech. Found. Div, ASCE, 94 (SM2), 371-389.

Hunt, C.H., Pestana, J.M., and Bray, J.D. (1998) "Effect of Pile Installation on the Shear Wave Velocity of a Soft Clay Deposit," Proc., The 5th Caltrans Seismic Research Workshop, Caltrans, Sacramento, CA, June 16-18, 1998.

Hunt, C.H., Pestana, J.M., Bray, J.D., Riemer, M.F. and Seed, R.B. (2000) "Geotechnical Field Measurements after Pile Installation in a Soft Clay Deposit," Geotechnical Engng Report No. UCB/GT/2000-05, Dep. of Civil & Environ. Engng., University of California, Berkeley.

Lehane, B.M. and Jardine, R.J. (1994) "Displacement-pile behavior in a soft marine clay," Canadian Geotechnical Journal, 31 (1), 181-191.

Lo, K.Y., and Stermac, A.G. (1965) "Induced Pore Pressures during Pile-Driving Operations," Proc., 6th Int. Conf. Soil Mech. Found. Engng., Montreal, Vol. II, pp. 285-289.

Nigbor, R.L., and Imai, T. (1994) "The Suspension P-S Velocity Logging Method," Geophysical Characterization of Sites, ed. Richard D. Woods (for the 13th Int. Conf. Soil Mech. and Found. Engng., New Delhi).

O'Neill, M.W., Hawkins, R.A. and Audibert, J.M.E. (1982) "Installation of pile group in overconsolidated clay," J. Geotech. Engng. Div., ASCE, 108 (GT11), 1369-1386.

O'Neill, M.W. (1999) Personal communication, also Terzaghi lecture, 1998

Orrje, O. and Broms, B. (1967) "Effects of pile driving on soil properties," J. Soil Mech. Found. Div., ASCE, 93 (SM5), 59-73.

Pestana, J.M. (1995) In Situ Testing of Soil Properties (after pile installation), Proc. 1993 Workshop/Seminar on Pile Installation Effects on Soil Properties, U.S. Corps of Engineers, Waterways Experiments Station, Vicksburg.

Radbruch, D.H. and Schlocker, J. (1958) "Engineering Geology of Islais Creek Basin, San Francisco, California." Miscellaneous Geologic Investigations, U.S. Geological Survey, Map I-264.

Randolph, M.F., Carter, J.P., and Wroth, C.P. (1979) "Driven piles in clay – the effects of installation and subsequent consolidation," Geotechnique, 29 (4), 361-393.

Reese, L.C., and Seed, H.B. (1955) "Pressure Distribution Along Friction Piles," Proceedings, Fifty-Eighth Annual Meeting of the American Society for Testing Materials (ASTM)

Roy, M., Blanchet, R., Tavenas, F. and La Rochelle, P. (1981) "Behavior of a sensitive clay during pile driving," Canadian Geotechnical Journal, 18 (2), 67-85.

Soares, M.M. and Dias, C.R.R. (1989) "Behavior of an instrumented pile in the Rio de Janeiro clay," Proc., 12th Int. Conf. Soil Mech. Found. Engng., Rio de Janeiro, Vol. I, 319-322.

Seed, H.B., and Reese, L.C. (1957) "The Action of Soft Clay Along Friction Piles," Transactions, American Society of Civil Engineers, Vol. 122, pp. 731-754.

Steenfelt, J.S., Randolph, M.F., Wroth, C.P. (1981) "Instrumented Model Piles Jacked into Clay," Proc., 10th Int. Conf. Soil Mech. Found. Engng., Stockholm, Vol. II, 857-864.

Stokoe, K.H., and Richart, F.E. (1973) "In-situ and laboratory shear wave velocities," Proc., 8th Int. Conf. Soil Mech. Found. Engng., Moscow, Vol. 1.2, 403-409.

Torstensson, B.A. (1973) "Friction piles driven in soft clay – A field study" Report R38:1973, Swedish Council for Building Research.

Vesic, A.S. (1972) "Expansion of Cavities in Infinite Soil Mass," J. Soil Mech. Found. Div., ASCE, 98 (SM3), 265-290.

DETERMINATION OF ROCK MASS MODULUS
FOR FOUNDATION DESIGN

By Brian D. Littlechild[1]FHKIE, M.ASCE, MICE, Stephen J. Hill[2] MHKIE,
Ian Statham[3] C.Geol., Glen D. Plumbridge[4] MSAICE, and Soon C. Lee[5] MHKIE

ABSTRACT

This paper reports on the ground investigation carried out for each test pile for a
series of nine 30MN full-scale load tests of piles founded on rock. The work was
carried out to provide data for the design of over 20km of the Kowloon-Canton
Railway Corporation West Rail Project Phase I by foundation contractors.

The ground investigation techniques used included Goodman Jack, High Pressure
Dilatometer, Self-boring Pressuremeter and Geophysics as well detailed logging of
the rock cores and laboratory testing for presentation in Rock Mass Rating (RMR)
format.   The advantages and limitations of each of the testing techniques are
discussed.

The paper concludes that the use of published correlations between stiffness and
rock mass rating parameters such as RMR and GSI would result in a fivefold over-
estimate of the stiffness. It is recommended that the stiffness of the rock may be
determined using conventional RMR data with the water and joint orientation
ratings set to zero. A new coefficient Modified Rock Mass Rating (RM^2) is
proposed for deep foundations on weathered rock.

INTRODUCTION

The Kowloon-Canton Railway Corporation (KCRC) proposed to construct the
30km West Rail Project Phase I in Hong Kong. The project required the
construction of 12.5 kilometres of viaduct, with seven of the nine stations, some
with development on top, supported on deep foundations. These structures were to

[1]Associate Director, Ove Arup & Partners 5th Floor, 80 Tat Chee Avenue, Kowloon Tong, HK
[2]Associate, Ove Arup & Partners 5th Floor, 80 Tat Chee Avenue, Kowloon Tong, HK
[3]Associate Director, Ove Arup & Partners 5th Floor, 80 Tat Chee Avenue, Kowloon Tong, HK
[4]Engineering Manager, Kowloon Canton Railway Corp., West Rail, Citylink Plaza, Shatin, HK

be founded on a wide range of rock types comprising: igneous (granite and granodiorite), volcanics (fine grained tuffs, rhyolites and pyroclastic breccia) and metasediments (siltstones, sandstones and marble). It would have been possible to follow the values of bearing capacity, laid down by the Hong Kong Building Authority. However, the regulations also permit the adoption of a 'rational' approach to foundation design. To adopt such an approach, when there was no precedent, required the designer to predict the settlement of the foundations on rock. To date there were no published values of stiffness from full-scale testing in Hong Kong that could be used even for the most commonly encountered rocks, granites and volcanics. So nine full-scale load tests were undertaken to provide design data for the eight rock types and for the different degrees of weathering encountered throughout the route.

In order to apply the results of the full-scale load tests to the whole of the route it was necessary to establish what combination of conventional coring, laboratory testing and in situ testing techniques could be used. Whilst a number of techniques were available, though not all in Hong Kong, they had never been used in conjunction with full-scale load tests to facilitate foundation design.

GROUND CONDITIONS

The 30km West Rail Phase I route crosses a large part of the New Territories in the Hong Kong Special Administrative Region, from Yen Chow Street just north of Kowloon in the extreme south, to Tuen Mun in the west as shown in Figure 1.

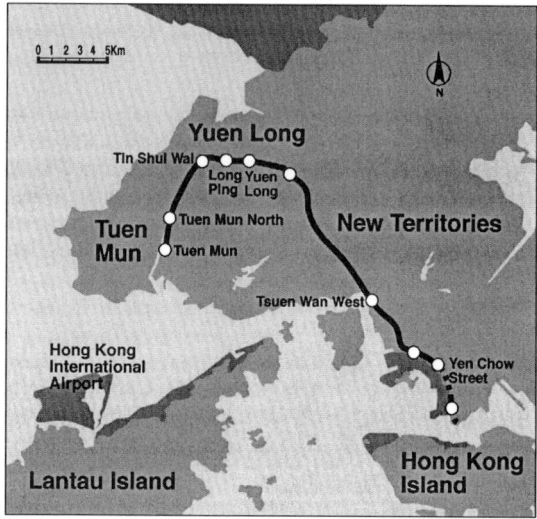

Figure 1 Route Plan and Test Sites

Much of the route is underlain by a deep sequence of superficial deposits, consisting of varying proportions of reclamation fill and soft to firm marine deposits and alluvium.

The solid geology is complex and highly variable. Most of the southern part of the route is underlain by granite and granodiorite. The granites are medium to coarse grained and massive, the weathering profile typically reaches 30m to 50m and locally may exceed 100m. Lateral variation in the depth of weathering is often rapid. The granodiorites are also coarse grained though sometimes often metamorphosed, heavily sheared and closely fractured. Again, deep weathering profiles with rapid lateral variation in depth may be present.

In the north and west the underlying bedrock consists mainly of metasediments (siltstones and sandstones), volcanics (fine grained tuffs and pyroclastic breccias) rhyolites and marble. The area is structurally complex with many thrust zones, intense deformation and shearing. Once again, deep weathering profiles characterize some of the rock types, especially the volcanics and metasediments. The marbles have been subject to solution, filled and unfilled cavities occur at depth in excess of 100m below ground level, some of which are more than 10m high.

GROUND INVESTIGATION TECHNIQUES

The following techniques were used to measure rock modulus directly within boreholes:

Goodman Jack (GJ): a 76mm diameter cylindrical jack designed to exert pressure on a borehole wall through two diametrically opposed steel platens, procedures and interpretation of the test is given by ASTM D4971-89 (1994). Displacement is monitored during three cycles of jacking the platens apart. Stiffness of the rock is calculated for the linear portion of the third cycle of the jack pressure displacement curve.

Borehole pressuremeters: a wide range of inflatable membrane pressuremeters have been developed for in situ modulus testing within a borehole, as discussed by Clayton et al (1995). Most of the tests were carried out in this study using the Cambridge High Pressure Dilatometer (HPD), developed for testing hard soils and weak rocks. At Yen Chow Street site there was a considerable thickness of completely weathered granite, in which the Cambridge Self-Boring Pressuremeter (SBP) was used (Wroth and Hughes (1973).

Geophysical technique: Geophysical methods offer the possibility of overcoming some of the inherent problems of direct borehole methods:

- the small volume of rock tested by borehole jacks and pressuremeters,
- the low sampling ratio,
- disturbance due to drilling,
- practical problems, especially in poorer quality rocks.

The method though is not widely used to determine in situ parameters for foundation design, and some reasons for this are discussed by Clayton et al (1995). In this study, the cross-hole seismic method was used as an indirect measure of the in situ modulus, as determined from ASTM (1995).

CLASSIFICATION SYSTEMS

Rock mass classification for the purpose of Presumed Allowable Bearing values in the design of deep foundations in Hong Kong are given by the Hong Kong Building Authority and published in PNAP141 (1995). Three categories of rock are defined, based on the weathering grade (Grades I to V from fresh to completely weathered), total core recovery in boreholes and a minimum unconfined compressive strength or point load strength index. While providing a conservative means of selecting bearing pressure for end bearing piles, the system does not lend itself to correlation with modulus since many rock mass factors known to be important are not included. Also, the use of total core recovery as a classification parameter is fraught with difficulty; drilling techniques and quality have an important bearing on recovery especially in very fractured rocks. Further, what is lost from the core may well be more important in rock mass performance than what is recovered.

Two principal rock mass classification systems are widely used in rock engineering, the Rock Mass Rating (RMR referred to as RMR_{76}) system published by Bieniawski and Orr (1976) and the 'Q'-system developed by Barton et al (1974). The 'Q'-system was originally developed for determining stand-up times and support systems in underground excavations. The use of RMR to design rock slopes and foundations as well as estimating the in situ modulus of deformation and rock mass strength is acknowledged by ASCE (1999). Both CIRIA (1995) and Clayton et al (1995) consider that of the well known classification systems RMR is more suited to foundation engineering purposes for piling than 'Q'-system as proposed by Barton (1983). For Hong Kong Geo (1996) recognises the RMR_{83} correlation with modulus given by Bieniawski (1979); and Serafim and Pereira (1983) for the deformation of axially loaded piles on rock. The RMR system uses factors to derive a numerical index, as shown in Table A1, given in the Appendix.

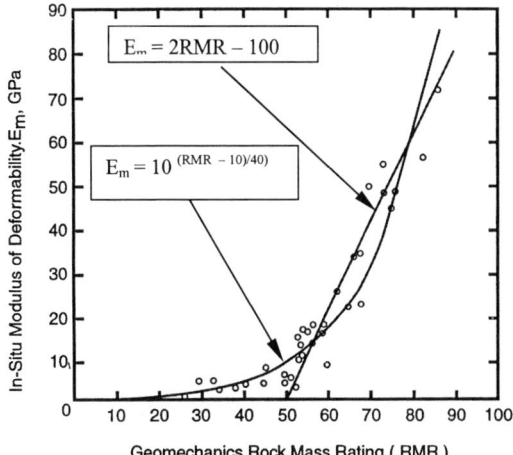

Figure 2 In Situ Modulus from RMR

Serafim and Pereira (1983) developed the Bieniawski (1979) RMR system for use in rock mass modulus determination where they increased the maximum water rating from 10 to 15 and as clarified in Bieniawski (1987) they also changed joint orientation weighting from -25 to 0. They analysed a number of case histories many of which involved dam foundations for which the deformation moduli were evaluated by back analysis of measured deformations. They proposed modifications to the RMR system to give a better fit at low values of modulus, as shown in Figure 2, and given by:

$$E_m = 10^{(RMR - 10)/40}$$

The original work by Bieniawski and Orr (1976) was based on an extensive review of large scale deformation modulus tests, including plates, tunnel relaxation, flat jack and pressure chamber tests, all of which were considered large enough to include the influence of discontinuities on rock mass modulus.

More recently Hoek (1994), proposed the Geological Strength Index (GSI) as a basis to estimate strength of jointed rock masses for underground excavations in rock. GSI is related to the Bieniawski and Orr (1976) classification system except that water is assigned a constant rating of 10 and joint orientation was set at 0 in all geological conditions.

A further modification to the work by Serafim and Pereira (1983), was proposed by Hoek and Brown (1997) for a system to determine modulus as given by:

$$E_m = [(\sigma_c /100)*10^{(GSI-10)/40}]^{\frac{1}{2}}$$

where σ_c is the uniaxial compressive strength of the intact rock pieces.

The Hoek and Brown (1997), modification is an empirical approach, applicable for $\sigma_c < 100$, where they consider that deformation of better quality rock masses is controlled by the discontinuities, while for poorer quality rock masses the deformation of the intact rock pieces contributes to the overall deformation process.

IN SITU MODULUS FROM GROUND INVESTIGATION TECHNIQUES

High Pressure Dilatometer (HPD) tests were scheduled as a means to provide data for the range of rock strengths that would not necessarily be covered by the Goodman Jack and Self-Boring Pressuremeter tests. However, the results of the HPD tests did not prove to be particularly successful. The moduli for the strong and massive rocks such as the metasiltstone and tuffs, exceeded the capacity of the HPD; typically 10GPa. In the highly fractured rock types such as the granodiorite (mostly RQD = 0), the fractured nature of the rock proved to be a hazard for the HPD membrane, and a number of membrane failures occurred. The results of the HPD tests have not therefore been considered further. The outcome is generally in line with the work of others (Haberfield and Johnstone (1993)).

Typical results for the Goodman Jack tests carried out in the moderately strong to strong granite and the moderately weak to moderately strong granodiorite are given in Figure 3. The tests were conducted on rock with a rock mass rating as given Table 1.

Table 1 Goodman Jack Test Results

Test Ref[1]	Rock Type	Depth (m)	Modulus (GPa)	GSI
TWW1	Granodiorite	33.0	3.8	52
YCS1	Granite	55.6	2.5	55
TMC1	Tuff	32.0	11	75
YUL1	Marble	40.4	18	76
TSW2	Metasiltstone	41.4	14	60

Note: 1 For test reference see Littlechild, Hill et al (2000).

The results presented in Figure 3 illustrate some of the problems considered to be associated with Goodman Jack testing in moderately weak to moderately strong rocks. Significant bedding of the platens is evident in the initial stage of the test. This effect is probably due to closing up of open fractures and stress relief from drilling.

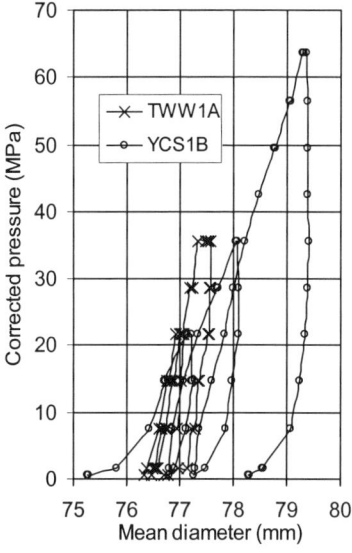

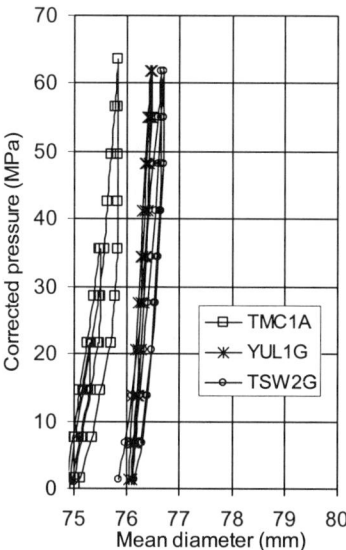

Figure 3 Goodman Tests in Granite Granodiorite

Figure 4 Goodman Jack Tests and in Tuff Marble and Metasiltstone

A further phenomenon evident from the test data for these rocks was the creep effects, during the hold period at the maximum jack load. This was particularly noticeable for the granodiorite rock when equilibrium was reached within 15 minutes for only the first stage.

The results for the strong to very strong rocks of the tuff, metasiltstone and marble are given in Figure 4. Movement of the jack platens, was smaller (less than 1mm) as compared to 3 to 4mm for the more fractured and weaker rock types in Figure 3. In these more massive and widely spaced jointed rocks, the creep effects were not observed.

Successful cross-hole geophysics testing was carried out in the metasiltstone, metasandstone, marble and granite and the results for modulus (E_o) are given in Figure 5, and presented in Table 2 with the rock mass rating. Cavities were encountered in the boreholes in the marble between 35m and 55m.

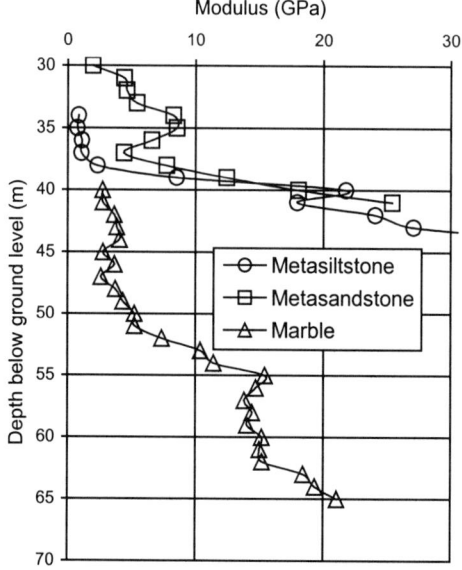

Figure 5 Cross-hole Tests in Metasiltstone, Metasandstone and Marble

Table 2 Cross-Hole Geophysics Test Results

Test Ref	Rock Type	Depth (m)	Modulus (GPa)	GSI
TSW1	Metasandstone	31 – 37	5-10	35 –57
TSW2	Metasiltstone	39.7 – 45.2	20-30	64 – 71
YUL1	Marble	53.0 – 65.0	10 – 20	71 – 84
YCS1	Granite	51.5 – 57.6	15 – 20	41 – 55

Figure 6 presents the SPT 'N' results carried out in the completely weathered granite (CDG). The usual practice in Hong Kong to evaluate stiffness in a completely to highly weathered soil, is given by Davies (1987), where

$$E = 1.4*N \ (MPa)$$

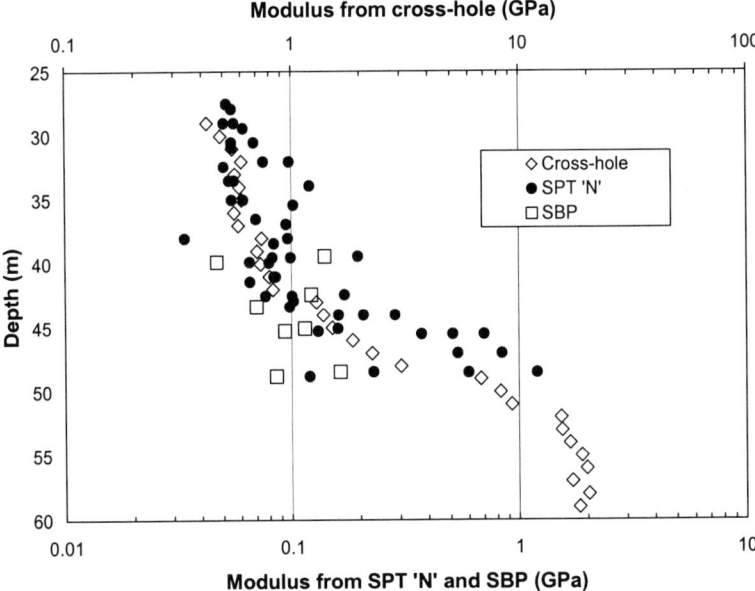

Figure 6 SPT 'N', Cross-hole and Self-Boring Pressuremeter Test Results in Completely to Highly Weathered

There is one order of magnitude increase in stiffness from the cross-hole results, when compared with the SPT 'N' approach, which is consistent with higher stiffness at small strain for the weathered rock as reported by Stroud (1988).

Results from the self-boring pressuremeter tests, which are marginally lower than the values derived from SPT 'N', are included in Figure 6 for completeness.

ROCK MASS MODULUS FROM PILE TESTS

Given in Table A2 in the Appendix is a detailed summary of the rock mass ratings for the individual factors for the rock cores and bearing stratum within two pile diameters at the individual test pile locations. The pyroclastic breccia is omitted as it is not appropriate to derive a value, because of the highly variable weathering grade in this rock type and particularly the presence of cavities after solution of the marble. The rock mass modulus for all other bearing strata has been evaluated for a pile toe settlement of no greater than 1% pile diameter.

$$\delta = Bq \, (1-\gamma^2)I/E$$

where $\gamma = 0.3$

More details including the performance of the end bearing pile test results are given by Littlechild, Hill et al (2000).

As discussed earlier the correlations between rock mass modulus and the rating systems proposed by Bieniawski and Orr (1976), Serafim and Pereira (1983) and Hoek and Brown (1994 and 1997) all assign different ratings for groundwater and joint orientation. However, having accepted that RMR_{76} and GSI are not exactly equal it is still useful to present the results from the testing carried out in Hong Kong on this basis in Figure 7 using GSI.

The data shows a general trend of increasing modulus with increasing GSI similar to that reported by Serafim and Pereira (1983) and Hoek and Brown (1997). However, the stiffnesses determined from the Hong Kong tests on tropically weathered strong rocks are about half an order of magnitude less than the values that would be derived from the tentative proposal by Hoek and Brown (1997). It should be noted that, when designing deep foundations, a rock mass modulus greater than 1GPa would normally result in tolerable settlements for building structures.

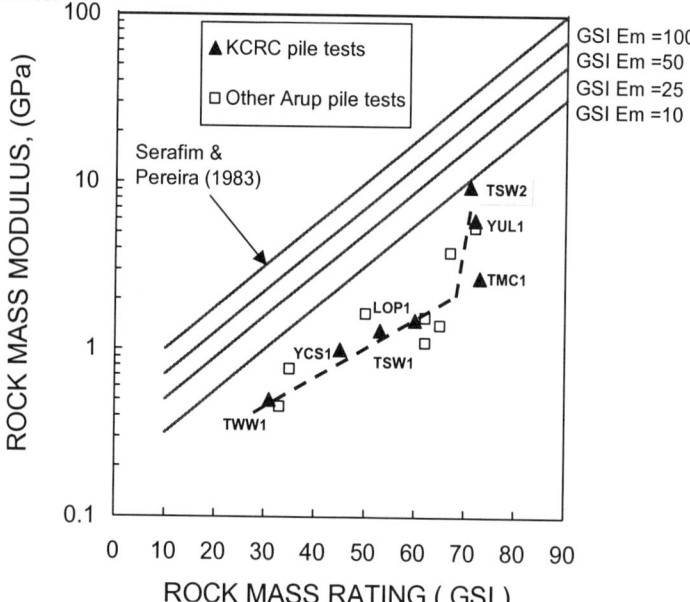

Figure 7 Rock Mass Modulus from Pile Tests and Other Systems

DISCUSSION

Rock Mass Rating and GSI: Given the mismatch between the data from the tests in Hong Kong and the GSI relationships proposed by Hoek and Brown (1997) a review is appropriate. The setting for the relationship they proposed is that of movements associated with large excavations presumably at depth within the rock mass. For such excavations the soft infill in the joints is more commonly fault gouge (see description of infill in Bieniawski and Orr (1976) and Hoek and Brown (1997)). Hoek (2000) states that though the relationship appears to work reasonably well more field evidence may result in the relationship being modified.

The testing reported in this paper was in quite a different geological and engineering framework. The stiffness of the rock has been determined for foundations a few metres across where the objective, in Hong Kong, was to minimise the depth of excavation into the rock, i.e. found close to rockhead. So in these circumstances the joint infill is typically preferential weathering of a jointed strong rock mass (fresh strength UCS 100 to 200MPa, degraded by weathering to 25MPa). It is considered that in these circumstances the deformation of the infill may be expected to predominate in the overall deformation process.

The authors accept Hoek and Brown (1997) recommendation that water and joint orientation do not affect stiffness and therefore consider that an appropriate ratings for water and joint orientation in deep foundation design are 0 for both. For this reason and to avoid confusion with other systems, the authors propose the term RM^2, (with a range +8 to +90) for the modified form of the rock mass rating using the same weightings as given by Bieniawski and Orr (1976) RMR_{76}, and adopted in GSI except that for water and joint orientation.

The moduli derived from the various ground investigation techniques, are presented in Figure 8 together with moduli from the pile tests. The HPD results have not been included for reasons discussed previously. Goodman Jack tests give significantly higher moduli compared to those derived from the pile test results. This is not surprising given the dimensions of the jack platen (approximately 200 by 75mm) compared to the pile dimensions (typically 1.2m diameter). Due to its size, the stiffness derived from Goodman Jack tests include the contribution from joints at close spacing within the rock mass. It is possible that tests on materials with RM^2 greater than 40 or so will be found to provide data that may be used for foundation design. However, again it should be noted that for foundations on rock with modulus greater than 1GPa the settlements would normally be tolerable for the design of building structures.

Figure 8 shows that the moduli derived from cross-hole are up to an order of magnitude greater than the stiffness mobilised by the foundation test piles. This is consistent with observations reported by Clayton et al (1995) where stiffness from seismic wave velocity may be applied directly to design though often, due to small strains, it overestimates stiffness when compared to back analyses of observed higher strain deformations. The results in the completely to highly weathered rock with SPT 'N' values between 20 and over 500 imply that the stiffness obtained is an order of magnitude higher than that almost universally used in geotechnical works in Hong Kong.

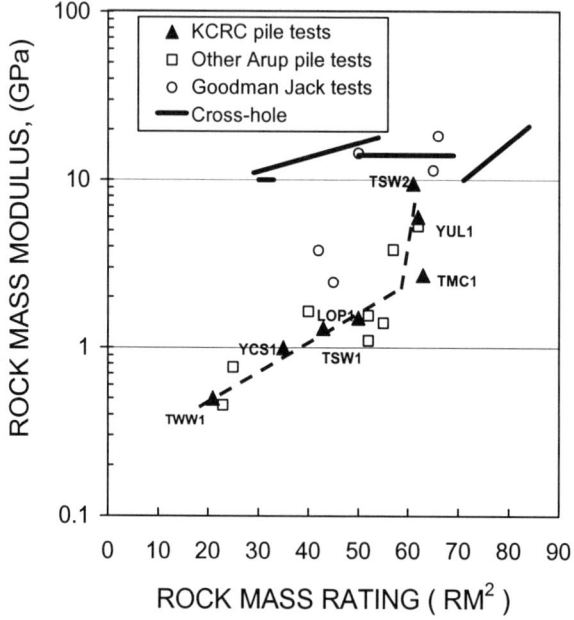

Figure 8 Rock Mass Modulus from Pile Tests and In Situ Tests

Goodman jack testing: It proved easy to carry out these tests in the more uniform and competent rocks with little fracturing or weathering. Undertaking tests in rock with RM^2 less than 30 to 35 requires careful consideration of the maximum pressure that will be applied in cycle 3 which the ASTM (1994) requires to be used to determine the modulus. If the maximum capacity of the equipment is assumed, as implied in the ASTM (1994), this may result in data that cannot be analyzed in accordance with the standard. ASTM (1994) does not recommend an alternative procedure though the authors would in future use lower pressures to see if meaningful results can be obtained.

Cross-hole testing: There are a number of limitations with this form of testing. It is necessary to reduce the background vibration noise to a minimum and so for this project the work had to be carried out at night. In addition, as some of the rocks were highly fractured the contractors, understandably so, would not allow the sonde to be lowered in the borehole, because of the implications of trying to recover the radioactive sonde. Each test required two boreholes and had to be lined with a UPVC lining to above the test zone. At higher values of RM^2, greater than 60, the may provide data that can be used directly in stiffness calculations where stiffnesses of greater than 10GPa are required. In light of the above comments the authors consider this test is not appropriate for routine design or quality control of site works.

Self-boring pressure meter testing: These tests could only be carried out in completely to highly weathered rock with N-values between 50 and a little in excess of 200. It was not possible to drill in the more competent highly or moderately weathered rock. The only other occasion when such tests have been undertaken in Hong Kong was reported by Malone et al (1997) for the back analysis of a diaphragm walled basement. The results were reasonably consistent with the assumptions for stiffness commonly adopted in Hong Kong based on SPT 'N' though, as was perhaps to be expected in weathered rock, there was a significant scatter in the results.

CONCLUSIONS

The testing undertaken for Kowloon-Canton Railway Corporation (KCRC) demonstrated a number of issues, for characterizing the bearing stratum and stiffness for deep foundations in Hong Kong:

- The authors consider the speed and simplicity of the Rock Mass Rating system, make it a practical method of assessing the rock mass.
- The results of the tests carried out in Hong Kong rocks, which include igneous, volcanic and metamorphosed sedimentary rocks indicate that even the lower bound correlation for weak rocks proposed by Hoek and Brown (1997) for rock excavations is not appropriate for foundation design on weathered rock.
- It is proposed that Bieniawski and Orr (1976) RMR ratings may be used as a design tool and for quality control on site provided it is adjusted by deleting the effect of water and joint orientation. To avoid confusion with earlier standard terms it is proposed that the term RM^2 is used to correlate rock mass quality with stiffness for foundation design.
- Insufficient Goodman Jack testing was done to establish whether it would be suitable design tool for foundations in weathered rock. It is possible that it would be appropriate for material with RM^2 greater than 40 though at values much higher than this the stiffness would be so high as to mean that there would in general be no need for such testing for foundations for buildings.
- It is considered that cross-hole geophysical testing is unlikely to provide an appropriate technique for the determination of foundation stiffness for normal piled foundations on moderately weathered to fresh rock in Hong Kong.
- The self-boring pressure meter test can only be carried out in highly weathered rock with an SPT 'N' value not a great deal more than 200.

- The high pressure dilatometer is not a suitable tool for measuring the stiffness of strong weathered rock in Hong Kong.
- All the test data is presented for completeness, and in the hope further testing and reporting in this form is published.

ACKNOWLEDGEMENTS

This paper is published with the permission of Kowloon-Canton Railway Corporation. The authors wish to thank the large number of Arup staff in particular Phil Dauncey who did much of the background work at the outset and the team who were involved in the site works and analysis of the results including James Sze, Nathan Wilmot, Richard Armitage, Tim Tham, Juliet Bird and Dominic Holt. Thanks are also extended to Hong Kong University of Science & Technology, in particular Professor Charles Ng and his students for Dora Nip, Terence Yau Lap and Jonathan Li. Recognition is due to the contractors Bachy Soletanche, EGS, Fugro Geotechnical Services, Gammon (HK), Lam Geotechnics and The Express Builders for their work in carrying out a series of unconventional testing works in a short contract period.

REFERENCES

ASCE, Rock Foundations, (1999) *Technical Engineering and Design Guides as Adapted from the US Army Corps of Engineers*, No 16.

ASTM (Re-approved 1994) Standard test method for determining the in situ modulus of deformation of rock using the diametrically loaded 76mm (3-in) Borehole Jack. *ASTM D4971-89.*

ASTM, (1995) Guide for planning and conducting borehole geophysical logging, *ASTM, D5753.*

Barton, N. Lien, R. and Lunde J. (1974) Engineering classification of rock masses for the design of tunnel support. *Norwegian Geotechnical Institute Publication 106.*

Barton, N. (July 1983) Application of Q-system and index tests to estimate shear strength and deformability of rockmasses. *Panel Report Theme II, International Symposium on Engineering Geology and Underground Construction,* Lisbon, Portugal.

Bieniawski, Z.T. and Orr C.M. (1976) Rapid site appraisal for dam foundations by the geomechanics classification, *Proc.* 12th *International Congress on Large Dams,* Mexico, Vol. 3.

Bieniawski, Z.T. (1979) The geomechanics classification in rock engineering applications, *Proc.* 4th *International Congress Rock Mechanics, ISRM,* Montreux, Vol. 2.

Bieniawski, Z.T. (1987) The rock mass rating system (Geomechanics Classification) in Engineering Practice. *Rock Classification Systems for Engineering Purposes. ASTM STP984.*

Clayton, C.R.I., Matthews, M.C., and Simons, N.E., (1995) *Site Investigation.* Blackwell Science Ltd, Second Edition.

CIRIA (December 1995) *Piled Foundations in Weak Rock.* Draft research report 509.

Davies, J.A. (1987) Groundwater control in the design and construction of a deep excavation, *Proceedings of the Ninth European Conference on Soil Mechanics and Foundation Engineering.*

Geo Publication No. 1/96, (June 1996) *Pile design and construction.* Geotechnical Engineering Office, Civil Engineering Department, Hong Kong.

Haberfield, C.M., and Johnston, I.W. (1993) Factors influencing the interpretation of pressuremeter tests in soft rock. *Proc. Symp. On Geotechnical Engineering of Hard Soils-Soft Rocks, Athens, Volume 1,* Balkema, Rotterdam,

Hoek, E. (1994) Strength of rock and rock masses, *ISRM News Journal.,2(2).*

Hoek, E. and Brown, E.T. (1997) Practical estimates of rock mass strength, *Int. J. Rock Mech. Min. Sci.,* Vol. 34 No. 8.

Hoek, E. (2000) Practical rock engineering, *www.rocscience.com/Hoekcorner.htm.*

Littlechild, B.D., Hill, S. J., Plumbridge, G.D. and Soon Chop L. (2000) Load Capacity of deep foundations on rock, *Proceedings New Technological and Design Developments in Deep Foundations, Geotechnical Special Publication, ASCE,* Denver.

Malone, A.W., Ng, C.W.W. and Pappin, J.W. (1997) Collapses and displacements of deep excavations in Hong Kong, *30th Anniversary Symposium of the Southeast Asian Geotechnical Society Deep Foundations*, Hong Kong,

PNAP141, (1995) Practice note for authorised persons and registered structural engineers, *Foundation Design-Building (Construction) Regulations 1990-PartVI,* Buildings Department, Hong Kong

Serafim, J.L. and Pereira, J.P. (1983) Considerations of the geomechanics classification of Bieniawski. *Proceedings of the International Symposium on Engineering Geology and Underground Construction,* Lisbon.

Stroud, M.A. (1988) The standard penetration test - it's application and interpretation. *Proceedings Conference Penetration Testing in the UK,* ICE, Birmingham.

Wroth, C.P. and Hughes, J.M.O. (1973) An instrument for the in situ measurement of the properties of soft clays. *Proc. 8th Int. Conf. Soil Mech. and Found. Eng.,* Moscow, Volume 1.2.

APPENDIX

Table A1 Summary of Bieniawski (1976) Classification Parameters, Ratings and Rock Mass Classes from Total Ratings

Factor	Range of Points		RMR	Qualitative Description	Class [2]
	Max	Min	$100-81$	Very good rock	I
Intact Rock Strength	15 (UCS > 200MPa)	0 (UCS < 1-3MPa)	$80-61$	Good rock	II
RQD	20 (>90%)	3 (<25%)	$60-41$	Fair rock	III
Joint Spacing	30 (>3m)	5 (<50mm)	$40-21$	Poor rock	IV
Joint Condition	25(very rough/no separation)	0 (soft gouge/open joints>5mm)	<20	Very poor rock	V
Groundwater [1]	10 (no flow)	0 (severe water problems)			
Joint Orientation [1]	0 (very favourable)	-25 (very unfavourable)			

Note: 1 Omitted from rating evaluation for deep foundations in RM^2
  2 Class given is not same as weathering grade used in Hong Kong

Table A2 RM^2 Evaluation for Bearing Stratum at Tests Pile Locations

Test Ref.	Depth (m)	Rock Strength Point Load (MPa)	Rock Strength UCS (MPa)	Rating	RQD %	Rating	Spacing of Joints (mm)	Rating	Condition of Joints	Rating	RM^2
TMC1	29.2-31.6	4.2	204	15	64	13	250	10	r,no sep,hd	25	63
TWW1	25.1-27.5	0.27 – 0.46	9	1	24	3	<50	5	r,sep sft	12	21
YCS1	49.3-51.7	0.2 – 0.6	22	2	43	8	<50	5	r,sep,hd	20	35
YUL1	40.6-43.0	1.3 – 3.8	55	7	95	20	<200	10	r,no sep,hd	25	62
LOP1	69.0-71.4	1.7	31	4	76	17	150	10	sm-sl,sep,sf	12	43
TSW1	40.3-42.7		76	7	58	13	100	10	r,sep,hd	20	50
TSW2	39.4-41.8	1-2.8	36	4	86	17	<350	20	sm, sep,hd	20	61

KARST INVESTIGATIONS – FOLDED OR FLAT, YOUNG OR OLD

By: Joseph A. Fischer[1], Member, Joseph J. Fischer[2] & Richard S. Ottoson[3]

ABSTRACT

All karst is not created equal. Variations in the age of the formation (e.g., Florida "limerock" versus the limestones of Wisconsin) result in strength variations and hence, performance. Variations in tectonic environment can yield flat carbonates (e.g., Kentucky's Mammoth Cave) versus the folded and faulted Appalachian carbonates (e.g., Virginia's Carlsbad Caverns) resulting in unusual patterns of solutioning as well as different failure mechanisms. Thus, geotechnical <u>investigation</u> and <u>analyses</u> must understand how the depositional and tectonic constraints of these different "rocks" can result in varying problems to investigators, analysts, designers and constructors.

This paper addresses the differences in possible structural failure or contamination modes that develop in these different rocks in relation to man's works. The study procedures presented are multi-phased; particularly if the geotechnical work is first used as input for planning purposes. A review of aerial photography, satellite imagery and available geologic information, as well as a reconnaissance to "ground-truth" the aerial photography and regional maps, are all generally part of the first stage of an investigation in karst areas. Subsequent phases consist of developing a geologic model of the site, then testing and refining it through field investigations and construction inspection. Direct investigation tools include test pits and borings using specific techniques, with perhaps the use of certain indirect (geophysical) tools to cautiously correlate between known hard data points.

The procedures are not unique, but represent an attempt to present economical investigative concepts that recognize the nature of different karst environments.

INTRODUCTION

Karst is found throughout the world. The term itself is the name of a province in what was once Yugoslavia. Technically, when used properly the term should be for the

---

[1] President, Geoscience Services, 25 Claremont Rd., Bernardsville, NJ 07924

[2] Vice-President, Geoscience Services, 25 Claremont Rd., Bernardsville, NJ 07924

[3] Associate, Geoscience Services, 25 Claremont Rd., Bernardsville, NJ 07924

pinnacled carbonate rock outcrops such as those found in the Karst region, not for any terrain overlying solutioned carbonates. The term limestone is often used generically to describe many carbonate "rocks" of varying composition, strength, susceptibility to solutioning, and subsurface morphology. In the United States karst is considered, very generally, to form in areas underlain by three types of rocks:

- The geologically recent (Cenozoic) sediments of the Caribbean, Bahama Platform and Florida;

- Older (Mesozoic and Paleozoic) limestones and dolomites such as those found in the folded and faulted valleys and ridges of the eastern and western United States as well as the flat-lying carbonates typical of the mid-continent; and

- The occasionally metamorphosed Proterozoic carbonates (marble).

Evaporites such as gypsum and salt, although often solutioned and perhaps "karstified", are not treated herein. However, many of the concepts presented herein can be applicable. The depositional environment of evaporites such as salt are different than limestones and dolomites thus, consideration must be given to their differing physical properties.

Unfortunately, most of the technical literature generated in the United States addresses the flat-lying rocks of the mid-continent (e.g., Figure 1) often reporting upon research in the Mammoth Cave area of Kentucky as well as the carbonate and salt layers of Ohio. However, not all carbonates are flat-lying or composed of hard, massively bedded rocks as one might presume from Figure 1 (the front cover of a conference proceedings). Florida and many Caribbean carbonates for example, may be flat-lying as shown on Figure 1, but are generally soft, sandy "limerock", not hard, crystalline materials. Appalachian

Figure 1 – Cover of karst proceedings.

Valley carbonates are generally hard, crystalline limestone and dolomites of the same age as their flat-bedded, mid-continent brethren, but are folded and fractured with seemingly always tilted bedding.

Thus, designing a suitable geotechnical investigation for structures located in carbonate terrane and correctly interpreting the results requires an appropriate geologic understanding of the nature of the karst in the locale of concern. All carbonates are not flat. All carbonates are not hard. Most carbonates do exhibit dissolution. Ground water movement may be slow to rapid, contaminant attenuation in their residual soils can be low to high, and structural failures occur in a variety of ways. Without attempting an extensive or complete characterization of potential failure modes, it may be useful to list what are the likely failure modes for the types of karst previously noted. Depending upon the focus of the investigation, "failure" can be related to either (or both) structural integrity and groundwater contamination.

A) Soft, recent karst is often characterized by "cover collapse" or "subsidence" sinkholes (e.g., Chen and Beck, 1989) where loose sandy materials fall or erode into generally near-vertical solution features within the underlying, relatively flat-lying "limerock". In Florida, for example, the permeability of the overburden is often high and both the primary and secondary porosity of the soft limerock can be large. Thus, ground water contamination can occur by pollutants or surface water rapidly entering the subsurface (see Figure 2C for example) and foundation failures can essentially occur by the loss of support under all or part of a structure.

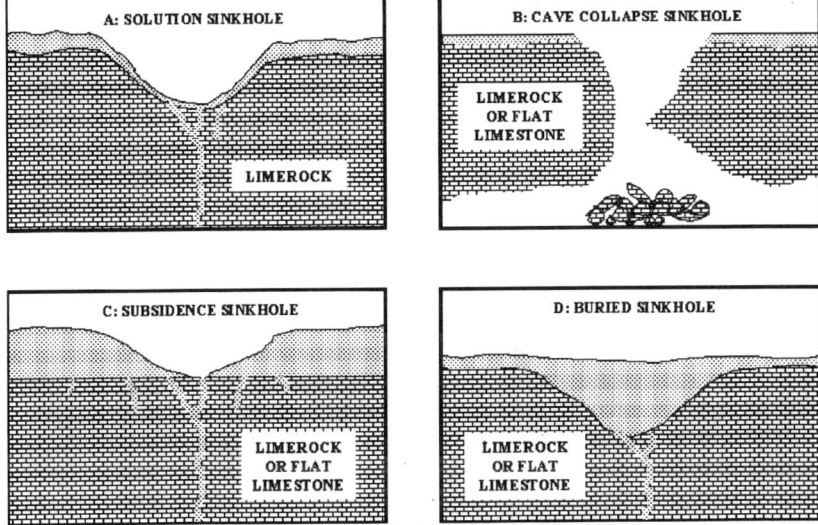

Figure 2 – Types of Sinkholes (Modified from Beck, 1991).

B) Structural failure within the flat-lying, but hard Paleozoic rocks of the central United States generally occurs through "cave collapse" (e.g., Fischer, et al, 1996). The roof of a linear solution feature (cave) that has formed generally along a bedding plane, may collapse as a result of continued solutioning or increased loading. Above the water table, virtually undiluted contaminants can travel for miles through open channels in these flat-lying rocks (e.g., Hannah, et al, 1989). See Figure 2B for example.

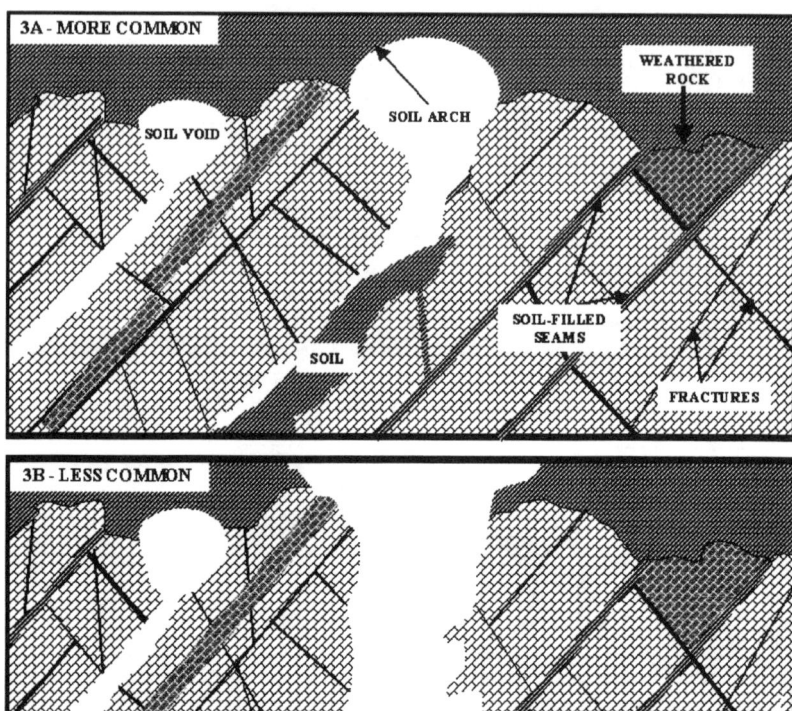

Figure 3 – Types of Sinkholes (Modified from Fischer, et al, 1996).

C) Conversely, a ravelling-type of failure generally occurs as a result of the erosion of overburden materials into solution features along inclined bedding or shear zones in, for example, the folded and faulted rocks of the Valley and Ridge Physiographic Province of the eastern United States (e.g., Fischer, et al, 1996). These sinkholes generally form from the rock surface upward and may not be evident until the soil arch (Figure 3A) collapses. Contaminants may not travel as far in open channels, but the aquifers are generally extensive and relatively unconfined. Also, wells are often relatively closely spaced in some areas of high population and wellhead protection considerations are quite important (e.g., Canace, et al, 1996). Foundation concerns are obvious as the arch above the soil void fails from overloading or roof material removal.

D) Solutioned cave features are often found in hard metamorphics of, for example, the western United States. Failure requires the overloading of a cavity roof or cutting into a cave (cave collapse sinkhole, see Figures 2B and 3B) rather than persistent dissolution or by erosion of the surficial soils into ancient cavities (Figures 2C or 3A).

E) A storm water concern in many rural areas of the country is sinkhole flooding. The closing of sinkhole throats by collapsing earth or filling during construction operations in addition to construction-related increases in runoff and/or temporary rises in the water table have created this problem in many areas.

Thus, Beck, 1991, in characterizing sinkhole occurrences (Figure 2) used flat-lying rock. Revising this assumption and eliminating the limerock-type failure yields Figure 3 for folded crystalline rocks (Fischer, et al, 1996).

These two-dimensional representations of failures yield a clue to what an investigator must consider in three dimensions. Horizontal variations in the flat-lying sediments are primarily due to the existence of near-vertical solution channels ("cutters") usually found within these rocks (e.g., Figures 2A, 2C and 2D). Cavities (or even caves) can extend for several to hundreds of meters laterally. Erosion occurs along exposed bedding and often along beds at different elevations as a result of ground water level changes that may have occurred over long time periods. Often, different rates and/or quantities of precipitation will change the elevations of such "conduit" flow phenomena as well as the volume of flow within a conduit.

For contrast, in folded rock terranes, significant strength and solubility variations can occur across bedding strike with, perhaps, down-dip continuity in physical properties. As fracturing can enhance permeability, variations of in situ stress magnitude and direction will further exacerbate the problem of trying to understand the variations in physical properties, performance and elevation across a folded and fractured, ancient Continental Margin area such as shown on Figure 4.

Therefore, in developing site investigation protocol, an understanding of the nature of the subsurface and its likely failure mode (or modes), as well as the more conventional engineering concerns of loading (both structural and contaminant) and grading (cuts and fills) must be considered. Often, the folded rock terrane (Figure 3A) and limerock terrane (Figures 2A, 2C and 2D) types of failures can be considered a soil mechanics constraint where a sinkhole or a zone of more compressible material results from the collapse of a soil void. In the more flat-lying, hard, crystalline rocks structural failures likely result from the collapse of cavity roofs - more of a rock mechanics failure. In addition, one must also recognize that variations in rock properties, including solutioning and weathering, will likely be more severe in a folded rock environment as a result of the numerous rock formations and/or beds that express themselves as the rock surface as one traverses the site (e.g., Figure 4).

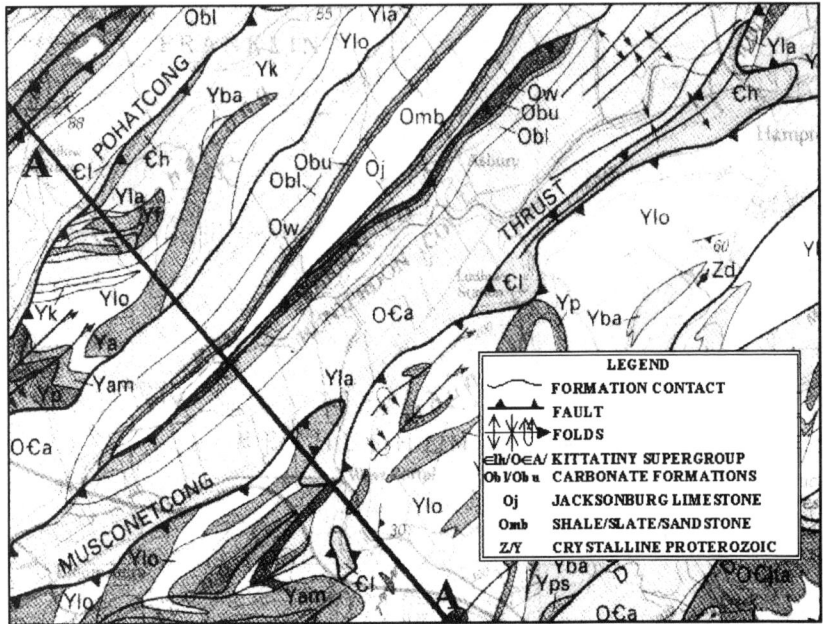

Figure 4 – Example of a "Continental Margin" site (from: Drake, et al, 1996).

The complexity shown on Figure 4 is not unusual in folded Appalachian terrane. The map shows a valley underlain by faulted Cambro-Ordovician-aged carbonates with the surrounding ridges and internal intrusives composed of Proterozoic-aged metamorphic rocks. The subsurface material variations along the cross-strike section A-A are enormous. The relatively small area between the Pohatcong and Musconetcong thrusts is microcosm of the Ridge and Valley system of the Mid-Atlantic states. Imagine the cross-strike variations in the anticlinal fold centered by the thin (white) "Omb" tongue crossing the central portion of Section A-A. The authors worked on a site lying across these Kittatinny Supergroup carbonates up to the thrust fault system at the Proterozoic rock boundary. Within each formation alone, the steeply dipping beds varied from massive to fissile, limestone to dolomite, solutioned to solid. Obviously, when designing a geotechnical study, one must recognize the problems inherent in investigating such a site.

Ground water contamination concerns will also vary as a result of the potential lack of chemical absorption, or the trapping of particulates, in undisturbed overburden materials, as well as rapid transport through open rock conduits common to all karstic terrane. Examples of such open channels are the vertical "cutters" of Figures 2A, 2C, and 2D, the fractures of Figures 3A and 3B, and the open cavities (or caves) typified in Figures 2B and 3B.

As a consequence of this variable nature and the type of hazard that ensues, standard soil mechanics investigation procedures are usually not effective in identifying the problems faced at a karst site. The basic geophysical tools such as reflection and refraction surveys, resistivity, and ground penetrating radar are often of marginal use. Tomography may present the best opportunity to reveal potential subsurface anomalies. The existence of highly irregular and variably weathered carbonate surfaces requires a level of geophysical resolution that can "see" soil voids or eroded channels as well as air, water and/or soft soil-filled rock cavities. The resolution necessary for conventional engineering investigations atop karst is much greater than normally attained in geophysical investigation. In addition, residual soils with high clay contents and varying moisture levels will limit radar penetration depths. Thus, while geophysical tools may be helpful in an auxiliary role (although often promoted as a cure-all), they are rarely as useful as conventional investigation tools unless one is looking for large, relatively shallow cavities (i.e., near-surface caves). Examples of geophysical tools, some of their advantages and some of their limitations are discussed in the Geophysical Investigation segment of the proceedings of the 1999 Karst Conference in Hershey, Pennsylvania (edited by Beck, Pettit and Herring).

It should be noted that the variable nature of the subsurface at a typical folded carbonate rock site that would generally limit the effectiveness of a geophysical procedure might not be present at all sites. Direct investigation (test borings, test pits and knowledgeable reconnaissance) prior to the establishment of a geophysical study can aid in making this determination as well as provide the hard data that is necessary to interpret the results of any geophysical survey.

As it relates to ground water concerns, conventional downhole permeability tests are often too specific for an overall understanding of geohydrological properties of a site's subsurface. Sometimes aquifer parameters can be developed from large scale well testing with numerous monitoring points, although conventional well modeling formulae are often inapplicable to an irregular, solutioned aquifer condition except as rough approximations with azimuthal variations. Solutioned carbonates are not isotropic, anisotropic, or slab-fissured/fractured rock as offered in many models. A finite element solution may be appropriate if the subsurface geometry could be defined, but at what level does the cost of measuring physical properties outweigh the usefulness of the model. Conduit flow is often assumed, but may not always be appropriate. Thus, approximations are usually always used in an attempt to estimate ground water movement rates and directions. In truly cavernous carbonates, dye tracing is a must to define rates, directions and even precipitation influence.

Virtually all subsurface investigations are driven by economics. Thus, it is necessary to understand that one cannot define all possible subsurface variations at most karst sites. From a practical standpoint, no geotechnically-oriented investigation will be funded at a level that allows a complete picture of the karst subsurface conditions. The authors believe that most investigation should be performed in phases and with a strong emphasis on obtaining general "geologic" data or approximate hydrologic parameters, rather than trying to define the actual physical parameters of the soils or rock at a particular location until one approaches the final design stage of a project. This multi-phased approach presumes that data will even be obtained during the construction phase and incorporated into final design and/or remediation. The use of a multi-phased approach allows a development/design team to become more familiar with the site subsurface conditions as the program progresses, adjusting design parameters and investigation goals to meet needs.

PHASE I - SITE AREA INVESTIGATION

In performing a study a great deal of information can be obtained at minimal costs in the initial (Phase I) stage of an investigation where a combination of published (and often unpublished) works by others, in conjunction with the investigators experience, is used. First a review is made of any data available from state and federal sources in the form of research reports and geologic maps which may describe the formations underlying the route/site and the extent of the folding and faulting that has affected them. Aerial photography, taken over a period of at least a decade, should be obtained as karst features may develop or change over time. Annual cycles of precipitation and drought can highlight different features. Features visible in early photographs may be obscured in later photographs by man's activities. Persistent lineaments and circular shapes are particularly suspicious when they are observed on a number of photographs taken over extended time periods. Photographs taken after periods of heavy precipitation can reveal interesting patterns of subsurface and near-surface water movement that can aid in developing a preliminary geologic model of the area in question. Low altitude, oblique aerial photographs taken at low sun angles and after precipitation events can be particularly effective in delineating areas requiring further on-site study.

Aerial photography, and sometimes satellite imagery, in combination with the available geologic information are used as the basis for a geologic reconnaissance. Simplistically, whatever the wavelength of the imagery, noting changes in the surface features that may reflect subsurface variations is useful. Areas that farmers avoid during cultivation often represent rock pinnacles or persistent sinkholes. Often, the farmers themselves are an excellent source of information regarding the location, size and persistence of sinkholes. Forested areas in otherwise fully farmed lands generally indicate areas of shallow rock or possibly intense sinkhole activity. Changes in vegetation sometimes indicate incipient sinkhole formation. Obviously, photo-lineaments, circular or linear depressions, possible sinkholes, disappearing streams, springs, wetlands etc. must be investigated on-site. One of the observations usually made in a geologic reconnaissance is that ground surface elevation variations across an area underlain by carbonate rocks is generally greater than they appear from a distance or as indicated on topographic mapping. These variations usually represent stratigraphic differences or geologic structure that has been emphasized by differential weathering.

A study of the well drilling information sometimes available from local and state sources can be of use. In the authors' experience, this data is usually sparse and lacking detail.

However, rock depth, water levels (considering that adjacent wells may have been logged years apart and/or during a different season) and especially water quality testing results, can be useful.

The amount of useful information that can be developed by an experienced engineering geologist/geological engineer from this preliminary phase is surprisingly large. The findings usually allow a number of decisions to be made for the subsequent phases of geotechnical investigation as well as for facility planning and preliminary design.

PHASE II - GEOTECHNICAL INVESTIGATION

Unless the results of the preliminary investigation frightens the owner/developer away, the needs of the proposed construction will likely lead to additional studies. A Phase II investigation is mostly used to confirm and expand Phase I findings through direct subsurface investigation which generally includes a number of test pits and borings at key locations. From a geohydrological standpoint, establishing ground water levels and monitoring water losses in test borings are generally all that is done at this time.

This phase of the investigation should be used to further explore geologically suspect areas identified in Phase I as well as critical facility areas and stormwater detention/retention areas. The emphasis of a Phase II investigation is to provide hard data and allow, at least conceptually, for structures, roadways, utilities and other facilities to be located and designed in a manner appropriate for the subsurface conditions likely to be encountered throughout the area of concern, not just at the locations drilled or excavated. However, so as not to escalate the cost of this phase, much extrapolation is necessary to expand the limited information gathered to this point into a loose but coherent geologic model of the entire route or facility site. One must also recognize that the variations that can occur over short distances at most karst sites will usually preclude the development of an all-encompassing geologic model. As an example, test boring and test pit data from a typical Appalachian Karst site is presented on Figure 5. The reader is encouraged to draw a cross-section with the pre-condition that drawing straight lines from data point to data point does not represent the true rock surface.

Test pits can be performed with conventional equipment and are obviously quite informative, particularly when a comparatively large area of rock is revealed in the pit bottom. The undulating or even disappearing rock surface is a surprise to many investigators and owners. Often, 3-meter long test pits have encountered shallow rock at one end and none at the other. Soil characteristics are also diagnostic. Is there a simple pattern of weathering in the pit side? Is there evidence of leaching or ground water movement? Is an old filled sinkhole or soil void revealed? Is the bedding dip identifiable in residual soils? Backfill in test pits should always be compacted to densities essentially the same as the in situ materials in an attempt to prevent the ponding of precipitation in the future and the enhanced opportunity to transmit water to the subsurface through loose soils. Very often, "soil logs" or geotechnical test pits become new sinkholes because poorly placed backfill allowed surface water to erode near-surface soils into an existing underlying rock cavity or open seam.

Test borings should be drilled using rotary wash drilling procedures without the addition of drilling "muds" to allow water loss depths and quantities to be monitored. Drill water is often lost just above the top of the rock indicating ground water flow had previously carried the generally clayey residual soils into a nearby (but not necessarily directly below) cavity within the bedrock. In the authors' experience, these water losses sometimes occur well above the rock (several meters) indicating a subsurface "throat" (often a precursor to soil void formation), a soil void (a sinkhole waiting to happen) or the gradual raveling of soils into underlying rock cavities. The soils encountered atop the rock in these areas of drilling water loss are usually soft, moist and clayey, or sometimes non-existent. Sampling is usually accomplished with a conventional split-spoon soil sampler, although the softer materials are sometimes difficult to retain even with the use of an appropriate sampler trap.

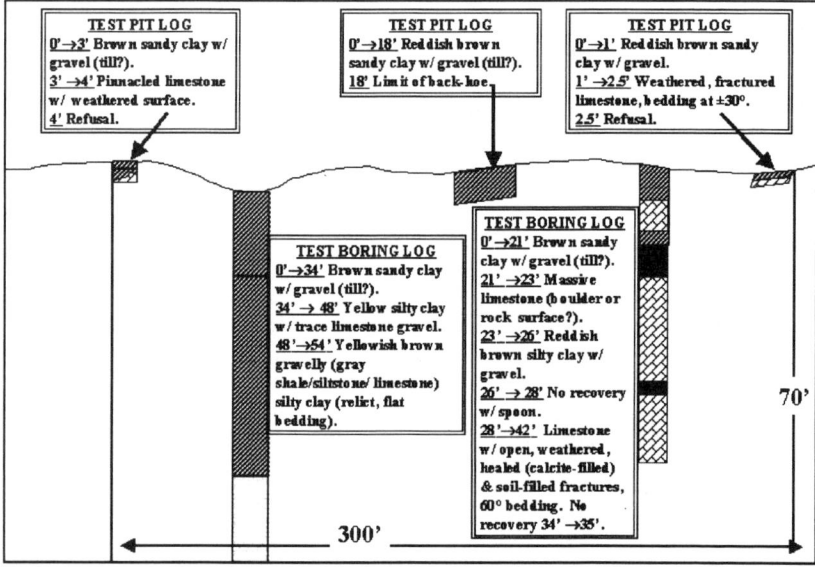

Figure 5 – Typical Appalachian karst site direct testing data.

Encountering rock is always an adventure. Is it sound, broken, weathered (to what degree), and what is the nature of the rock surface? Numerous core barrels, roller bits and drill rods remain in the ground as testimony to the difficulty of advancing a hole in creviced, solutioned or pinnacled limestone. If "rock" is encountered, is it a pinnacle, boulder, or a sound bed underlain by a series of weathered saprolites? Did a thrust fault push a more competent dolomite over more solutioned limestone, or did fault or glacial movement remove all the soft weathered materials?

Rock core should be extracted with a double- or triple-tube, split core barrel. This equipment takes no longer to advance in the borehole and perhaps only an extra few minutes to extract the core. The information gleaned from a core run lying in the split barrel, not "hammered" out of a conventional core barrel, is invaluable. Numerous fractures, some open, some soil-filled, are in evidence. Often, the firmer cavity fill materials are recovered. Three to six meter runs with a 1.5-meter core barrel are not uncommon as the core barrel encounters voids or soft materials that are washed away during drilling. Making a 1.5-meter core run through competent material and obtaining 100% recovery with high Rock Quality Designations (RQD) is rare in many karst terranes. Obviously, a driller with extensive experience in karst is very important as one usually cannot sample (or retrieve a sample from) every variation and rarely has drilling fluid return throughout the entire borehole depth. This requires careful observation of the changes in drilling advance and rig noise; something the driller is usually better at interpreting than most "inspectors".

Many indirect procedures have been advanced for the detection of subsurface cavities and rock contours that have not yet become surface features (depressions, outcrops or dolines). A host of geophysical techniques have been suggested as suitable including seismic reflection and refraction, electrical resistivity and conductivity, self potential, ground penetrating radar, and gravity surveys (see, for example, Beck, et al, 1999). If one is aware of the nature of the subsurface target and the resolution available with most commercial equipment using conventional procedures, it is difficult to visualize how these conventional geophysical prospecting procedures can be considered any more than marginally useful even in areas of flat-lying rock. They function best in an auxiliary role to the performance of direct measurements of subsurface parameters. Pinnacled rock, large floating boulders, the size and nature of soil voids, clayey soils, and the lack of a coherent ground water table compromise the use of geophysical surveys in most karst engineering applications.

It is valuable to recognize, however, that at times such indirect methods can be effectively combined with direct methods of investigation such as test borings and test pits. Tomography may be useful, albeit expensive. However, geophysical methods cannot be substituted for carefully drilled test borings, qualified full-time inspection, experienced and professional drillers, and large diameter (NX-sized or larger) double- or triple-tube split barrel coring. There is invaluable information to be gained through examining water-softened soils immediately above the rock surface, noting drilling water losses at various depths, observing clay-filled seams within the rock, transported soils, stained joints, and/or weathered cavity or joint sides in a core run laid out in the core barrel, or watching the drill rods fall through a 3-meter void and into soft cavity fills.

Pneumatic probes can be quite valuable when used in conjunction with test borings because of an air-track's mobility, speed of drilling and relative economy, at least at shallow depths (less than about 12 meters). Although no core is extracted, rock depths and qualitative competency can be estimated. With an experienced operator, it is possible to see or feel major changes in the stiffness of the overburden soils as well as by measuring penetration rate differences while drilling in variable quality rock. Noting the depths that air loss is experienced also aids in understanding the subsurface. Observing air returns from a previously drilled probe hole after noting its disappearance from the hole in progress can be quite informative. Unfortunately as with test boring water losses, the lack of air return precludes even looking at the returned samples ("cuttings") of the subsurface.

Aside from the soil and rock data available from test borings and pneumatic probes, grouting (or attempting to grout) the borehole is informative as grout-take is monitored for depth and volume, then compared with the borehole volume. A conventional cement, water, bentonite "slurry" can be used and the quantities of grout placed downhole should be measured. However, after several cubic meters of grout have been introduced into a 0.1 cubic meter borehole, it is often necessary to employ unconventional means to plug the hole so as to prevent water infiltration and possible future sinkhole occurrence.

PHASE III - FOUNDATION OR GEOHYDROLOGICAL INVESTIGATION

Phase III utilizes the subsurface data and resulting geologic model developed from the work to this point, considers the proposed facility layout, and performs more detailed subsurface investigations (as necessary) to further expand the information in critical areas. The investigation can be performed in a single field program or just prior to planned, phased construction. In either case, all subsequent work should always build upon the previous work. For example, if construction is planned for an area identified as likely having shallow rock, relatively closely spaced test pits may be appropriate to help in estimating whether rock will be encountered and whether it is "rippable" or will require some other means of removal (e.g., blasting or "hydrohammer"). Signs of past incipient sinkhole formation may be observed from the ground surface in a test pit wall. If highly sensitive or heavily loaded facilities are included, then drilling below the proposed locations or even at each column location may be warranted. Probe hole depths will depend upon column loads, potential load distribution configuration, type of subsurface (flat or folded rock), strength of the rock, depth to rock, judgement, and experience.

Geohydrological concerns are not as easily addressed. How do you characterize an almost infinitely variable medium? Should one consider conduit flow, fracture flow, matrix flow, or is it a combination of all of them? The answer is probably yes, but what was the question? In the flat rock scenario, one can easily visualize flow through a series of conduits, perhaps formed at different elevations within the rock mass as the water levels change during long-term river "downcutting". Are there enough fractures to merely connect these semi-horizontal conduits or are the fractures dense enough to control flow or yield a fractured block-type of subsurface model? Perhaps in geologically recent sediments, an isotropic model or a simple(?) anisotropic media can be used to characterize a carbonate aquifer. Again, one must consider the long-term effects of the potential slow dissolution of fractured materials or the slow precipitation of minerals into fractures. In addition, one must consider what might be the effect of short-term erosion of the weathered materials on the deposition of surficial soils into any underlying cavities.

Hence, in planning a phased geohydrologic investigation, one must use the initial phases to establish a rational geologic model before attempting to estimate suitable parameters. This is obviously a difficult task.

What assumptions must one work with when planning a Phase III geohydrological investigation? Is the investigation related to water supply or contaminant concerns?

In a sandy limerock aquifer, it may be feasible to consider Darcy flow equivalents, but anisotropy is certainly not impossible. With a high water table in Florida for example, conduit flow may not be of great significance unless ground water levels are lowered considerably by pumping.

Conversely, the more flat-lying limestones often exhibit conduit flow, but through which conduit and during what kind of recharge event? Also consider the difficulty in attempting to use test and observation wells to monitor water flows. How do you hit the conduit(s) of significance? Often, dye tracing and monitoring springs and streams is the procedure of choice in flat rock areas when one wishes to find where it goes and at what concentrations (Hannah, et al, 1989). A specialist in dye tracing, receptor design and evaluation is a necessary adjunct should there be a need for dye tracing.

Aquifer tests using a relatively large number of observation wells may provide a key to water supply and contaminant movement studies in for example, folded and fractured eastern United States rocks. Hopefully, the rocks are fractured enough that some form of conventional(?) Darcy flow continuum model is applicable. However, eastern United States tectonism has certainly left its impact on these rocks and anisotropy (as a minimum) virtually always exists.

Sometimes Phase III investigations can include remedial grouting (e.g., Fischer and Fischer, 1995). An exploration/grouting program is generally quite cost-effective at this stage (prior to construction), rather than after failure has occurred. Although remediation can be accomplished in a number of ways, downhole tremie grouting with a cement-based slurry or grout and shallow excavation with sinkhole (rock) throat closure are generally the most economical.

A version of dynamic compaction can also be quite effective in remediating larger areas or along a roadway. This procedure entails dropping large weights from great heights to:
1. Collapse any near-surface cavities and/or voids;
2. Seal cavities with overburden soils, and
3. Densify the soils overlying competent rock.
Shallow rock voids may be exposed by this process and remediated by filling them with an appropriate cement slurry or grout.

SUMMARY AND CONCLUSION

Karst terrane is widespread, but not necessarily well understood from an engineering design and construction standpoint. Misunderstanding karst concerns can be partially attributed to the lack of appreciation of failure types and subsurface variability considering the proposed projects uses. Thus, adequately performing an appropriate geotechnical or geohydrological investigation that provides satisfactory answers to karst-related design, contamination and construction problems within normal project economic limitations requires a multi-faceted study with a great deal of extrapolation from, hopefully, representative data.

One must understand the nature of the subsurface in relation to design and construction concerns. The great variations in subsurface conditions over short horizontal distances in karst make it economically impractical to completely characterize its physical nature. An investigation will likely not be able to completely define the subsurface in the area of concern. Both direct and indirect procedures for investigation can be useful, but only certain direct procedures are effective in identifying the actual subsurface conditions, and only at the test pit or test boring location. Each boring samples only a small vertical segment. Hence, extrapolations of these data are more difficult at a karst site than in most geologic environments. Be aware of the failure mechanisms likely at your site and be careful.

REFERENCES

Beck, B.F., 1991, On Calculating the Risk of Sinkhole Collapse, *Appalachian Karst*, Proc. of the Appalachian Karst Symp., Nat. Speleological Soc. Redford, VA.

Beck, B.F., A.J. Pettit, and J.G.Herring (Editors), 1999, *Hydrogeology and Engineering Geology of Sinkholes and Karst – 1999*. A.A. Balkema, Boston, MA.

Canace, R., D. Monteverde, and M. Serfes, 1996, Karst Hydrogeology of the Shuster Pond Area, Hardwick Township, Warren County, New Jersey, *Karst Geology of New Jersey and Vicinity*, Proc. of 13th Annual Mtg. of Geological Assoc. of NJ, Whippany, NJ.

Chen, J. and B.F. Beck, 1989, Qualitative Modeling of the Cover-Collapse Process, *Engineering & Environmental Impacts of Sinkholes & Karst*, Proc. of 3rd Multidisciplinary Conf. on Sinkholes and the Engineering and Environmental Impacts of Karst, A.A. Balkema, Boston, MA.

Dilameter, R.R. and S.C. Csallany (Editors), 1977, *Hydrologic Problems in Karst Regions*, Western Kentucky Univ., Bowling Green, KY.

Drake, A.A, Jr., R.A. Volkert, D.H. Monteverde, G.C. Herman, H.F. Houghton, R.A. Parker, and R.F. Dalton, 1996, *Bedrock Geology Map of Northern New Jersey*, USGS Map I-2540-A.

Fischer, J.A. and R.W. Greene, 1984, New Jersey Sinkholes - Distribution, Formation, Effects, & Geotechnical Engineering, *in Proc. of 1st Multidisciplinary Conf., on Sinkholes*, A.A. Balkema, Orlando, FL.

Fischer, J.A., R.W. Greene, J.J. Fischer, and F.W. Gregory, 1982, Exploration Grouting in Cambro-Ordovician Karst, in *Grouting, Soil Improvement and Geosynthetics*, v. 1, ASCE Geotech. Publ. No. 30

Fischer, J.A. and J.J. Fischer, 1995, Karst Site Remediation Grouting, *Karst Geohazards*, A.A. Balkema, Rotterdam.

Fischer, J.A., J.J. Fischer, and R.F. Dalton, 1996. Karst Site Investigations: New Jersey and Pennsylvania Sinkhole Formation and Its Influence on Site Investigation, *Karst Geology of New Jersey and Vicinity*, Proc. of 13th Annual Mtg. of Geological Assoc. of NJ, Whippany, NJ.

Fischer, J.A., J.J. Fischer and R.J. Canace, 1997, Geotechnical Constraints and Remediation in Karst Terrane, *Proc. of the 32nd Symposium on Engineering Geology and Geotechnical Engineering*, Boise, ID.

Fischer, J.A., J.J. Fischer and R.S. Ottoson, 1999, Design of Geotechnical Fabrics for Septic Systems in Karst, *Hydrogeology and Engineering Geology of Sinkholes and Karst – 1999*, A.A. Balkema, Rotterdam.

Hannah, E.D., T.E. Pride, A.E. Ogden, and R. Paylor, 1989, Assessing Ground Water Flow Paths from Pollution Sources in the Karst of Putnam County, Tennessee, *Engineering & Environmental Impacts of Sinkholes & Karst*, Proc. of 3rd Multidisciplinary Conf. on Sinkholes and the Engineering and Environmental Impacts of Karst, A.A. Balkema, Boston, MA.

Reitz, H.M. and D.S. Eskridge, 1977, Construction Methods Which Recognize the Mechanics of Sinkhole Development. In *Hydrologic Problems in Karst Regions*, Western Kentucky Univ., KY.

Sowers, G.F., 1999, *Building on Sinkholes*, ASCE, NY, NY.

Subject Index

Page number refers to the first page of paper

Author Index

Page number refers to the first page of paper